T0302299

High Speed Rail in the U.S.

High Speed Rail in the U.S.

Super Trains for the Millennium

Edited by

Thomas Lynch

Center for Economic Forecasting and Analysis
Florida State University, Tallahassee

Gordon and Breach Science Publishers

Australia • Canada • China • France • Germany • India • Japan • Luxembourg • Malaysia • The Netherlands • Russia • Singapore • Switzerland • Thailand

Amsteldijk 166
1st Floor
1079 LH Amsterdam
The Netherlands

British Library Cataloguing in Publication Data

High speed rail in the U.S. : Super trains for the millennium
1. High speed ground transportation – United States
2. Railroad engineering
I. Lynch, Thomas
385′.0973

ISBN 90-5699-606-1

This book is dedicated to my children *Thomas Averry Koontz Lynch* and *Katelyn Koontz Lynch* and their generation as a gift and a guide with hope for a better future.

It is also dedicated to my parents *Thomas Amos Lynch* and *Winifred Marie Louise Caron Lynch* and their generation in thanks for the gifts they have given my generation—gifts of freedom, dedication, intelligence, choice, example and love.

CONTENTS

FIGURES

TABLES

INTRODUCTION

The nineteenth century witnessed the emergence and domination of railroads as they crisscrossed our nation and integrated our frontier economy. The early part of the twentieth century saw the emergence and eventual domination of the automobile and airline transportation systems as they eclipsed the nation's railroad network. These newer modes of transportation were more able to effectively link time and space and efficiently move the high-end goods, services and passengers of our rapidly changing economy, as well as enhance capital and labor productivity.

The latter part of the twentieth century has also experienced accelerated emergence of a new and equally powerful transportation force reaching across all of Europe and much of developing Asia—but leaving the United States behind. High speed rail (HSR) systems began operating in Japan in 1964 and France in 1981. These systems have spread to all the developed nations of Europe and are making rapid inroads in Taiwan, Korea and most of developing Asia.

To date, billions of passengers have traveled in both parts of the world with virtually no fatalities. Hundreds of millions of passengers arrive on time more efficiently than any other transportation mode currently in operation can boast. The economies of these regions of the world are increasingly integrated and dominated by these modes and will continue to be so well into the new millennium. This proven record is propelling Europe towards even bolder plans for binding the continent together in the future.

This HSR ground transport web will integrate the continent's diverse economies. Today, European nations have constructed over 2,500 miles of HSR service across the continent. Prospects are growing to integrate the entire continent of Europe with a ribbon of 5,600 miles of very high speed rail (155 mph plus) and an additional 9,300 miles of upgraded track capable of speeds in excess of 124 miles per hour.

With this HSR system in place, the European continent, with an estimated population of 385 million, is united and constitutes the single largest integrated market in the developed world. HSR will serve as a vital backbone for accelerating this mega-economy into one of the principal global competitors in the increasingly competitive global market. It will help define the European economy as one integrated whole. There is active discussion about running HSR systems across Eastern Europe and throughout much of western Russia, as those nations' economies emerge in the early years of the new millennium. Further R&D in Japan and Europe is likely to extend rail speeds to 210 miles per hour within five years. Some researchers predict that Maglev systems may extend the speed range of high speed ground transport (HSGT) to almost 300 miles per hour by 2010. The U.S. cannot afford to be left behind.

The U.S. has lagged in development of HSR for a variety of reasons: lack of institutional support from government and private sectors; lack of high-quality technical information for policy makers and the general public; and powerful and sustained opposition from existing entrenched modes with well-established (and publicly subsidized) market niche share of goods and passenger movements.

HSR service has "nudged" the door open in the Northeast Corridor with successful deployment of the mid-range speed Metroliner service. The service levels offered by the Metroliner are limited by corridor constraints (alignments not designed for high-speed service) and conflicting regional commuter rail and freight movements sharing the same alignment. Even with these constraints Metroliner service in the Washington–New York corridor is a commercial success. Metroliner carries 40% of the commercial travelers in this corridor and transports more passengers than all the airlines combined. Where HSR exists among appropriate city-pairs, this trend will accelerate well into the twenty-first century and be a boon to those regional economies.

As unabated population growth and urban density increase across the east and west coasts, in the midwest and across the booming metropolises of the south, demand for more efficient transportation systems is inevitable. The expansion of automobile peak gridlock from morning and evening rush-hour phenomena to a day-long grind and increasing wing-lock and airport capacity constraints (resulting in billions of dollars of airliner delay costs) across the major U.S. urbanized areas will accelerate this demand. These pressures and the emergence of a unique American private-public partnership will revitalize (revolutionize) financing and operation of these systems in the future. Once the efficiency and quality of true HSR services are proven in one region of the U.S., other regions will demand equivalent access and service.

This book is intended to help fill some of the technical and policy information gaps identified earlier as sources constraining further development of HSR systems in the U.S. Each author selected to contribute to this HSR "handbook" is a recognized expert and leader within his respective field of research. Each was chosen and developed his contribution to this text to be understood by the interested "non-expert." The book is written and organized to address several key areas of HSR development and will provide a foundation for both well informed and novice with a single reference on the key aspects of planning, development and implementation of high speed rail systems in the U.S. To this end, this text can help fill the information void and provide high-quality public information for policy makers, investors, students, public agency staff, environmentalists and the general public.

Numerous feasibility studies and other professional and academic publications have been written on a variety of technical HSR system issues. The contributors to this volume believe that providing a single comprehensive reference of this kind will help further the informed debate on the merits of rapid deployment of this new technology in the U.S. This in turn will hasten the emergence of HSR to a place of prominence as one of several pivotal transportation investments needed as a complement to other modes of travel. It should serve as an important multi-modal link at the center of U.S. transportation policy agendas well into the new millennium and help the U.S. remain a vital and competitive global economic force.

1 HIGH SPEED GROUND TRANSPORT : OVERVIEW OF THE TECHNOLOGIES

Tony R. Eastham,
Professor,
Department Civil and Electrical Engineering
Hongkong University of
Science and Technology
Clearwater Bay, Kowloon Hong Kong.

1.1. INTRODUCTION

High speed ground transportation (HSGT) are those systems capable of sustaining operating speeds over 200 km/h. The age of HSGT began in 1964 with the opening of the Shinkansen between Tokyo and Osaka. Over the past 20 years, a host of systems have been developed and introduced for speeds up to 300 km/h in Europe. Further R&D in Japan and Europe is likely to extend rail speeds to 350 km/h within five years. And Maglev systems may extend the speed range of HSGT to 500 km/h by 2010.

Unlike Europe and Japan, America is struggling with HSGT. It can reasonably be claimed that the concepts of superconductive electrodynamic Maglev originated in the U.S.. While much credible early research was conducted in America, it is evident that leadership was lost to Japan by the early 1970's. By 1975, Maglev R&D in the U.S. was virtually non-existent. There was a brief renaissance in the late 1980's and early 1990's with the National Maglev Initiative. While a number of interesting ideas emerged, nothing was built, and Maglev in America has now returned to its previous minimal state, with a few advocates promoting projects based on undeveloped technology. In America, HSGT is not yet viewed as an opportunity to reshape society and to enhance mobility, but is still being put to the limiting tests of benefit/cost ratio and return-on-investment in a competitive transportation market place. But maybe there is hope with the realization that an integrated multimodal national transportation system is needed in the 21st century.

The author offers his perspective on the past history, current status, and likely future directions for development of high speed ground transportation.

1.2. EVOLUTION OF TRAINS

The train has long been a vital transportation mode in North America, and indeed in all industrialized countries throughout the world. Trains were historically important for their role in linking the population centers on the east and west coasts of continental North America and for opening up vast tracts of land in the mid-west and south for development in the late 19th century.

Before the second world war, trains were the dominant mode, and an extensive network of lines across the continent moved people, food stocks, primary resources and industrial products. However, with the post-war development of the U.S. interstate highway system and an extensive network of airways, railroads declined in significance and all but the mainlines have been abandoned. Most people in North America choose to travel by road for short trips and by air for longer distances. Remaining interurban and commuter rail services now account for less than 2% of passenger-km per year. Rail still moves substantial freight, but even here trucks have become the dominant carrier.

In the face of declining market share, it is difficult for North America to justify a major investment in HSGT and to claim any technological leadership in trains. The situation is quite different in other countries. Steel wheel-on-rail trains are now operating at speeds up to 300 km/h, and Maglev vehicles are being developed and tested for introduction at speeds of 400–500 km/h, perhaps within ten years.

It is convenient to define HSGT as those guided systems which are capable of sustaining operating speeds in excess of 200 km/h (125 mph). By this definition, only one system in North America makes the cut – Metroliner service from New York to Washington, D.C. While this train covers the 360 km trip in a scheduled time of 2 hours and 45 minutes, for an average speed of 131 km/h (82 mph), cruising speeds of 210 km/h are achieved over certain track sections. Federal Railway Administration regulations limit operating speeds to 216 km/h because of safety considerations related to the shared right-of-way (with freight and lower speed passenger trains) and the signaling system.

By contrast, HSGT has been widely implemented and has carried almost 4 billion passengers without a single fatality in countries outside the U.S.

In Japan, the famed Tokaido Shinkansen (bullet-train) opened between Tokyo and Osaka in 1964. Over the years, operating speeds have been raised from 210 to 270 km/h, and trip times for the 515 km trip from Tokyo to Osaka have been cut from 4 hours to 2 hours 30 minutes. Japan now has 1836 km of Shinkansen double-track in service, from Morioka in northern Honshu to Hakata in Kyushu. These lines carry over 275 million passengers per year. Active development programs are being pursued by the regional railway companies, and a STAR 21 (Super Train for the Advanced Railway of the 21st century) test train has achieved a speed of 425 km/h.

France has the fastest train service in the world – the Train à Grande Vitesse (TGV). The TGV Atlantique and Nord lines have a maximum speed of 300 km/h. Paris forms the hub of a TGV network that extends north to Lille and the Channel Tunnel, west to Tours and Lemans, and south to Lyon. TGV trains operate to the Mediterranean coast of France, into Belgium (Brussels) and Switzerland (Geneva), and have been introduced into service in Spain between Madrid and Seville. TGV service is planned for implementation in Korea between Seoul and Pusan.

Germany also has a high speed train, the Inter-City Express (ICE), which is now providing 250 km/h service between Hannover and Wurzburg and from Mannheim to Stuttgart. This system, like those in Japan and France, operates on a dedicated right-of-way. Passenger safety is assured by advanced control of train movements and no high speed at-grade crossings. Sweden has adopted a somewhat different approach by building a train, the X2000, that makes best use of existing railroad infrastructure by actively tilting the passenger compartment relative to the wheeled bogies to avoid subjecting the passengers to uncomfortable lateral forces while negotiating curves at high speed. The X2000 achieves a top speed of 210 km/h (to be raised to 250 km/h by 1998) on the 456 km line from Stockholm to Gothenborg. The ETR-450 tilt-body train provides similar service between Rome and Firenze in Italy.

Thus the technology of high speed trains has advanced and is providing invaluable passenger services in both Europe and Asia. In these countries, it represents the technological evolution of ground transportation infrastructure for societies that have never relied upon the automobile for mobility and which have not needed to develop an extensive network of airways for intermediate distance travel.

And what of the situation in North America? Public interest has remained high, but progress towards implementation of high speed rail has been frustratingly slow. There is recognition that the continent's mobility is being threatened by gridlock on arterial freeways in metropolitan areas and by so-called winglock at hub airports at peak times. A renaissance for rail is being proposed. Millions of dollars have been spent in technological evaluations, system design studies, route surveys and ridership assessments for corridors such as Pittsburgh-Philadelphia, Las Vegas-Los Angeles, San Francisco-Los Angeles-San Diego, Dallas-Houston-San Antonio, Miami-Orlando-Tampa, and Toronto-Ottawa-Montreal. Little progress has been made because the economics are marginal, and because of the reluctance of federal and state governments and the private sector to commit substantial funds for implementation.

However AMTRAK, the U.S. rail passenger carrier, is not standing still. Its flagship operation over the electrified line from Washington, D.C. to New York runs at speeds up to 210 km/h with Metroliner service, and carries more passengers than either of the air shuttles on this route. In the spring of 1996, AMTRAK plans to issue a phased $600 million contract to purchase 26 high speed train sets for use in the North-East corridor. These will eventually offer service at speeds up to 225 km/h in the Boston-New York-Washington, D.C. corridor once upgrading of the railroad infrastructure (electrification and route re-alignment to remove some speed-limiting curves) is complete. These high-speed train sets will be manufactured in the U.S., likely as a joint venture with an off-shore developer. Companies competing for the AMTRAK contract are offering technologies based on a tilting version of the TGV, on the X2000, and on an integration of the ICE with Fiat tilting technology.

In the future, we may be able to travel at speeds up to 500 km/h in Maglev vehicles. Maglev is the generic name for a family of technologies in which the vehicle is suspended, guided and propelled by means of non-contact magnetic forces. While concepts for such systems can be traced back to the early 1900's, the age of Maglev was borne in the mid-1960's. Following the pioneering work of two physicists, Jim Powell and Gordon Danby, working at Brookhaven National Laboratory on Long Island, N.Y., the U.S., Japan, Germany, U.K. and Canada all began R&D programs. Those in Japan and Germany matured to the development, testing and pre-commercial demonstration of vehicle systems that could be operational by 2005–2010.

It is evident that there is a spectrum of technologies which either can or will soon be able to serve intercity corridors with HSGT in the speed range of 200–500 km/h.

1.3. HIGH SPEED TRAIN SYSTEMS

HSGT systems can be broadly categorized into three speed ranges :

200–250 km/h
- Diesel electric trains
  (including British HST)
- Electrified tilt-body trains
  (including Swedish X2000, Italian Pendolino, Spanish Talgo)
- Electrified non-tilting trains
  (including U.S. Metroliner)

250–350 km/h
- Electrified non-tilting trains
  (including Japanese Shinkansen, French TGV, German ICE)

400–500 km/h
- Magnetically suspended vehicles
  attraction (electromagnetic) mode ; German Transrapid
  repulsion (electrodynamic) mode ; Japanese Linear Express

The gap from 350 – 400 km/h could be filled by further advances in the technology of electrified high speed rail over the next decade.

Some of the characteristics of high speed systems are listed in Tables 1.1 through 1.3 below, and photographs of a number of trainsets and Maglev vehicles are presented as Figures 1.1 through 1.6.

Table 1.1 presents features of two operational tilt-body trains. Tilting systems make it possible to increase train speeds without modifying existing curved trackage while not exposing passengers to uncomfortable lateral accelerations. The train body control system acquires information from accelerometers at the front of the train and computes the required angle of inclination. This angle signal is used to control hydraulic actuators which tilt each car body on the train relative to its bogies by the appropriate angle related to train speed and local track curvature.

The ETR-450 was designed and built by Fiat Ferroviaria and is operated by Italian State Railways. This train entered revenue service in early 1988 between Firenze and Rome at a maximum speed of 250 km/h. The second generation ETR-460 was introduced in 1994.

TABLE 1.1. Features of 250 km/h Tilt-body Trains.

CHARACTERISTIC	*ABB X-2000*	*FIAT ETR-460*
In commercial service	1990	1994
Top speed	276 kph	300 kph
Service speed	210 kph	250 kph
Vehicle type	locomotive-hauled with driving trailer	EMU
Consist	1.4-DT (in service)	M-T-M-M-T-M-M-T-M (9 cars)
Seating capacity	200 (all 1st class); 254 mixed	456 + 2 disabled
Propulsion	ac 3-phase asynchronous; 815 kW; 4 powered axles	dc; 500 kW, body-mounted; 12 powered axles/train
Braking	Blended regenerative, discs, magnetic rail brake	Blended rheostatic and discs
Power supply	OCS 15 kV, 16 2/3 Hz single phase	OCS 3 kV dc
Axle load	18.25 tonnes (max.)	12.5 tonnes
Unsprung mass	1.8 tonnes/axle	1.6 tonnes/axle
Maximum tilt	8°	8°
Other features	Steerable powered trucks	Partially active lateral suspension

The X2000 train Fig 1.1 was designed and built by Asea Brown Boveri (ABB) and is operated by Swedish State Railways. The train entered revenue service on the Stockholm-Gothenborg line in September 1990 and operations have been expanded as trainsets are delivered. X2000 operates at a maximum of 210 km/h in Sweden but has been tested to 276 km/h in Germany. An X2000 trainset has also

Figure 1.1. The X2000 tilt-body train rounding a cure on the Stockholm-Gothenborg line in Sweden.

been tested and demonstrated in the U.S. North East corridor, and was operated very successfully in revenue service between New York and Washington, D.C. at speeds up to 225 km/h during spring 1993. Similar service demonstrations have been conducted with ICE and Talgo equipment.

Table 1.2 presents features of the highest speed rail systems in the world: TGV in France and Spain Fig. 1.2, ICE in Germany Fig. 1.3, and Shinkansen in Japan Fig. 1.4. All of these trains are non-tilting with ac drives and operate on dedicated rights-of-way of limited curvature (typically a minimum radius of 6000 m on high speed sections), with infrastructure that is maintained to allow sustained high speed operations.

High speed rail systems have and will continue to evolve to take advantage of beneficial advances in a number of technologies, including microelectronics in controllers and signaling, power electronics and devices in propulsion equipment, advanced materials in car body construction, and aerodynamic shaping for energy efficiency and noise control. The train systems shown in Table 1.2 represent recently introduced high speed rail equipment in France, Germany and Japan, respectively.

The Train à Grande Vitesse (TGV) is built by the GEC-Alsthom consortium and is operated by SNCF. The Sud-Est line from Paris to Lyon was opened for revenue service in 1981 and operates at a top speed of 270 km/h. The TGV-Atlantique trains operate on the more recently built line to the west of Paris to Tours and Lemans. At 300 km/h, TGV-A equipment provides the fastest revenue service in the world, commencing operations in September 1989. Prior to being opened for revenue service, a specially equipped and shortened TGV-A trainset (1.3-1) was tested at speeds up to 515.3 km/h; a world speed record for any rail vehicle. The Paris-Lille section of the TGV-Nord was opened for revenue service, also at 300 km/h, in 1993 and extension to Brussels is under construction.

The Intercity Express (ICE) was designed and built by a Siemens-led consortium and is operated by German Federal Railways. ICE entered revenue service in June 1991 on a new line from Hannover to

TABLE 1.2. Features of High Speed Trains with 300 km/h Capability

CHARACTERISTIC	*TGV-Atlantique*	*ICE-1*	*SERIES 300 (Nozomi)*
In commercial service	1989	1991	1992
Top speed	515.3 kph (1.3-1)	406.9 kph	325.7 kph
Service speed	300 kph	250 kph; 280 kph on some track segments	270 kph
Vehicle type	Articulated trainset	Loco-hauled	EMU
Consist	1.10-1	1.13-1 or 1.14-1	16:5(M-T-M) and cab car
Seating capacity	369 coach; 116 first	681 (1.14-1)	1,323
Propulsion	ac synchronous, 1100 kW, 8 axles powered	ac asynchronous, 1200 kW, 8 axles powered	ac asynchronous, 300 kW, 40 axles powered
Braking	Blended rheostatic, disc and tread brakes	Blended regenerative and discs	Blended regenerative, disc and eddy-current
Power supply	OCS 2 x 25 kV, 50Hz	OCS 15 kV, 16 2/3Hz	OCS 2 x 25 kV 60Hz
Axle load	17 tonnes	20 tonnes	11.3 tonnes
Unsprung mass/axle	2.2 tonnes	1.87 tonnes	1.86 tonnes

Figure 1.2. An AVE-TGV train on the Madrid-Cordoba-Seville line in Spain.

Figure 1.3. An ICE train on the Hannover-Wurzburg line in Germany.

Würzburg. The ICE fleet is presently limited to 250 km/h by alignment geometry and superelevation compromises required to allow shared use of track with other equipment. Operating speed can be raised to 280 km/h over certain sections of track to recapture lost time. Second generation ICE trainsets (2/2) are now in production.

The first Shinkansen service was opened in 1964 between Tokyo and Osaka–in time for the Olympic Games of that year. The line was subsequently extended to Okayama, Hiroshima and Fukuoka by 1975. Additional lines from Tokyo to Morioka and from Tokyo to Niigata were opened in stages from 1982 to 1991. Shinkansen trains are currently restricted to 270 km/h by noise and vibration concerns (by local communities) and by the alignment geometry of the track. The new Series 300 Shinkansen equipment has achieved 325 km/h under test conditions. It entered revenue service with JR Central in mid-1992 and the fastest train (Nozomi) has cut travel time from Tokyo to Osaka (515 km) to 2 hours 30 minutes.

1.4. MAGLEV SYSTEMS

High speed rail represents the evolution towards its ultimate operational performance capability of a technology which began with Stevenson's Rocket in England in the early nineteenth century and which became a workhorse of the industrializing world for the transportation of raw materials, products and passengers. Operating speeds have continuously increased over time and it now appears that revenue service at 320 km/h and perhaps 350 km/h is technically feasible and achievable.

Maglev has been developed in the belief that very high speeds, in the 400-500 km/h range, are desirable to achieve trip times that are competitive with short-haul air travel, to relieve winglock at major airports and gridlock on freeways in metropolitan areas, and to meet the mobility needs of tomorrow's post-industrial society in an environmentally sustainable manner.

Maglev is a revolutionary form of transportation which has been researched, developed, tested and brought to the stage of pre-commercial demonstration over a period of 30 years. It uses magnetic forces for the non-contact suspension, guidance and propulsion of vehicles at speeds up to 500 km/h. Table 1.3 presents the features of the two superspeed Maglev systems; the German Transrapid (Figure 1.5) and the Japanese Linear Express (Figure 1.6). Note that both HSGT systems are at least 10 years away from revenue service. Most of the features in Table 1.3 therefore refer to planned operational vehicles.

TABLE 1.3. Features of Superspeed MAGLEV Systems

CHARACTERISTICS	*EMS SystemTransrapid*	*EDS System JR Linear Express*
Country of origin	Germany	Japan
Status	Pre-deployment testing	Development testing
Geometry	Up to 10% gradient, 5800m radius curve	Up to 4% gradient, 8000m radius curve
Guideway	2.8m wide, simply-supported guideway, steel or concrete	2.8m U-shaped concrete guideway, simply supported
Power supply	20kV, 3-phase VVVF to windings in guideway	VVVF inverters feeding windings in guideway
Control and communications	Unique ATC/ATO with moving block; VHF vehicle-wayside communications	Under development, but similar in principle to that used by Transrapid
Key guideway features	Guideway carries windings for iron-core LSM guidance rails, waveguide. Required aligment tolerances +/– 0.6mm for stator packs	Guideway alignment tolerances are less critical than for EMS system due to larger air gap [100-150 mm vs. 8-10 mm]; air-core LSM
Vehicle type	Articulated EMU	Articulated EMU
Dimensions (l x w x h)	25.5m x 3.7m x 3.95m	21.6m x 2.8m x 2.85m
Consist size standard	2, 4, or 6	14
Capacity standard	200, 400, 600	988
Propulsion	Iron-core LSM	Air-core LSM
Braking	LSM thrust reversal; eddy-current emergency brakes	LSM thrust reversal; aerodynamic emergency brakes; aircraft discs on undercarriage wheels
Guidance	Non-contact magnetic attraction	Non-contact magnetic repulsion
Body structure	Aluminum alloy	Aluminum alloy
Suspension	Magnetic primary; pneumatic secondary	Magnetic primary; spring secondary
Axle load	1.6 tonnes/m	1.0 tonnes/m
Design speed	400–450 km/h	500 km/h
Hotel power collection	Non-contact linear generator	Non-contact linear generator
Noise level	84-86 dB(A)	N.A.
Key operational features	Propulsion, braking are not adhesion-limited	Larger air gap; inherently stable suspension; faster

Figure 1.4. A Shinkansen train near Fukushima station on the JR Tohoku line.

Figure 1.5. A TRANSRAPID 06 vehicle on the Emsland test track in northwest Germany.

The Linear Express, as being developed and tested (in 1997) at full scale by Japanese Railways, is magnetically levitated by electrodynamic suspension (EDS) by means of the repulsive force generated between superconductive magnets carried by the vehicle moving adja-

Figure 1.6. An artist's rendition of a Linear Express Maglev vehicle in Japan.

cent to and inducing current in discrete short-circuited aluminum coils mounted along the guideway. The EDS system is characterized by a guideway clearance of 100-150 mm at high speed. However, the levitation force is speed dependent and the vehicle requires wheels at low speed. An EDS Maglev vehicle must achieve a forward speed of 60–120 km/h (dependent on design) to generate magnetic levitation, above which its wheels may be retracted. EDS is dynamically stable but underdamped. Passive damping in the primary suspension and a secondary suspension are required to achieve good ride quality.

TRANSRAPID, as being developed and tested by Magnetschnellbahn GmbH, is magnetically suspended by electromagnetic suspension (EMS) by means of the attractive force between vehicle-borne iron-cored electromagnets and ferromagnetic guideway components. This mode of suspension is inherently unstable and must be dynamically stabilized by active feedback control of the magnet excitation in response to changes in the gap. The suspension gap is 10–15 mm, i.e. an order of magnitude less than that for EDS, but is nearly speed independent. An EMS vehicle therefore does not need wheels. EMS technology has also been developed by HSST Corporation in Japan for intermediate and low speed applications, and has been implemented as a low speed shuttle for the Birmingham Airport People Mover in England.

Both EDS and EMS vehicles have magnetic guidance, whereby a lateral displacement generates a strong restoring force towards the Centre of the guideway. The Linear Express uses "null-flux" guidance produced by cross-coupled coils mounted on each side of the guideway. Transrapid uses a separate set of controlled electromagnets carried by the vehicle and interacting with ferromagnetic rails on the sides of the guideway structure.

Maglev vehicles require a non-contact means of propulsion and braking which is compatible with the operating clearance of the magnetic suspension. The EDS Linear Express uses an air-cored linear synchronous motor (LSM). Three-phase excitation of armature coils produces a magnetic wave into which an array of on-board superconductive magnets is locked. This same magnet array is used for levitation and guidance. The speed of the magnetic wave is determined by armature input frequency, providing precise speed control of the vehicle with high power factor-efficiency operation. As the vehicle moves along the guideway, successive coil groups are powered up and the vacated sections are shut down. The EMS Transrapid system uses an iron-cored linear synchronous motor, using the suspension electromagnets for excitation. The principle and mode of operation is the same as for the Linear Express vehicle. A major advantage of the LSM is that propulsion power is not transferred to the vehicle and processed on board—the guideway armature is the high power component of the motor. Hotel power can be transferred to the vehicle by non-contact transformer effect, with on-board batteries for back-up purposes.

In Japan, the Linear Express concept was developed through a series of test vehicles until by 1979 a speed of 517 km/h had been reached on the 7 km test track near Miyazaki in Kyushu. Development has continued to the construction of a 41.8 km pre-commercial vehicle test and demonstration facility in Yamanashi prefecture near Tokyo. The first 18.4 km of this double track guideway will allow essentially all aspects of an operational system to be tested, starting in 1997, including full scale vehicles passing at full speed (500 km/h) in a tunnel. It is anticipated that 500 km/h Maglev could be ready for deployment on a new line between Tokyo and Osaka by 2007.

In Germany, the attraction mode electromagnetic "TRANSRAPID" Maglev system has been under development for high speed ground transportation since the late 1960s. Evolutionary development through a series of test vehicles led to the construction of the Emsland facility in the early 1980s. This 31.5 km figure-of-eight shaped guideway allows full scale vehicles to be tested and demonstrated under close to operational conditions. The pre-production vehicle TR-07 has been under evaluation for almost five years, and has now shown itself to be ready for implementation at speeds of 400–450 km/h. This technology has been selected by the German government for a new line from Hamburg to Berlin by the middle of the first decade of the 21st century to enhance east-west travel links following the reunification of Germany.

Maglev has had a prolonged adolescence. Unlike high speed rail, it is a revolutionary rather than evolutionary technology. It is likely to find application only where high passenger utilization can justify the cost. In the U.S., Maglev R&D was stalled after a brief period of research activity in the early 1970's at Ford Motor Company, the Stanford Research Institute and the Massachusetts Institute of Technology in parallel with that in Germany and Japan. However, American Maglev was rejuvenated in the late 1980's as a result of entrepreneurial and political hustling , and the government-sponsored National Maglev Initiative (NMI) was initiated as an attempt to utilize a number of relevant advanced technologies (cryogenics, high temperature superconductivity, power electronics, aerodynamics, control, and vehicle dynamics) from the aerospace and related industries to initiate development of a second generation Maglev system to meet the needs and conditions of North America. This generated a lot of interest and activity, a number of interesting ideas, but little follow-through. Again, Maglev R & D activity declined to a low level in North America as a consequence of the completion of the NMI program, the limited federal commitment to on-going R & D, and a reluctance by the private sector to commit to long term Maglev development. By 1994, Maglev R&D in North America had returned to its previous minimal state, with a few entrepreneurs promoting undeveloped technology for various projects.

Twenty five years ago, R & D in Maglev was justified by the perceived need for ground transportation at 450–500 km/h to provide intercity trip times in high population density corridors up to 600 km in length that are competitive with air travel, amid concerns about the cost and availability of oil-based fuels. At that time, the practical speed limit for steel wheel on rail was thought (by many) to be about 250 km/h. However, as we have seen, high speed rail has continued to be developed, as researchers have gained a better understanding of wheel-rail dynamics, aerodynamics, power pick-up by pantograph from a single-phase catenary, and operating speeds have reached 300 km/h. Even more impressively, a high speed train (a shortened specially-equipped TGV consist) has been tested at speeds up to 515.3 km/h. While it is not claimed to be feasible to run a passenger train at this speed, 350 km/h is considered to be technically and operationally possible. The margin of speed advantage for Maglev is therefore being cut, and it remains to be seen whether any government is prepared to proceed with investment for the implementation of a radically new ground transportation technology for a rather limited number of very high speed lines when proven advanced high speed rail can offer trip times that exceed potential Maglev capability by only 20–60 minutes for a typical intercity trip (depending on Maglev speed) at significantly lower capital and operating cost.

1.5. COUNTRY DEVELOPMENTS

The operational status and development plans for HSGT in a number of countries around the world will now be briefly reviewed.

1.5.1 Europe

The countries of the European Union (EU) are planning to network their high speed rail lines in an incremental manner over the next 25–30 years. The near-term goal builds on the existing high speed routes to form some regional networks. Routes that were operational in 1995 are:

Hannover-Wurzburg (328 km)
Madrid-Cordoba-Seville (472 km)
Mannheim-Stuttgart (130 km)
Paris-Lyon (427 km)
Paris-Lemans/Tours (202 / 224 km)
Paris-Lille (227 km)
Rome-Firenza (318 km)
Stockholm-Gothenborg (456 km)

The mid-term plan supplements the existing routes to form a mid-European network with the perimeter of London-Glasgow-Hamburg-Munich-Marseilles-Lisbon-Bordeaux, but is still poorly connected to Italy and Scandinavia.

The long-range vision would form a truly continental network. Such a network requires 9,000 km of high speed lines and 10,000 km of upgraded lines. With the addition of 11,000 km of link and feeder lines, this configuration has 30,000 km of rail line throughout Europe, representing an investment of over 100 billion ECU. To quote from a publication of the Community of European Railways (1989), "a high speed rail network which is energy-efficient, environmentally-friendly, economical and technically advanced, will reshape the transport scene. It can help to resolve the worsening congestion problems in air and road travel. It will bring fast, reasonably-priced and comfortable travel to the people of Europe. It will also provide a unique opportunity for regional, social and economic development within the EU. It represents a powerful catalyst for European integration".

France

Following the opening of the Paris Sud-Est and Atlantique TGV lines in 1981 and 1989, respectively, the TGV Nord is being extended from Paris to Brussels . The Paris-Lille section is now operational, and speeds up to 320 km/h are planned for revenue service within two years. Intermodality has been enhanced by the opening of a TGV station at Ch. de Gaulle Airport. Meanwhile, following completion of the Channel Tunnel, Eurostar service was introduced in 1994 to provide a direct rail link between Paris and London. Near-term plans include the extension of the Sud-Est and Atlantique lines to Marseille and Montpelier and to Rennes and Angers, respectively. Longer term plans include the construction of a high speed network of approximately 7,000 km of line with 220 km/h capability, with Paris as the primary node.

Germany

ICE service is now operating on new lines from Hannover-Würzburg and from Mannheim-Stuttgart and on a network of older trunk lines in Germany. Near term-plans include a Cologne-Frankfurt link. The German Infrastructure Plan makes provisions for a 4,500 km network of high speed line, with 800 km of new infrastructure (including the ICE lines already constructed). The Transrapid Maglev system has been undergoing development and testing on the 31.5 km Emsland loop near Bremen for the last 10 years. The current 07 vehicle is a pre-deployment prototype. Planning is underway to implement Transrapid between Hamburg and Berlin by 2005.

Italy

Rome-Firenze Directissima service became operational in 1990, using ETR-450 equipment. Plans include the completion of a T-shaped "alta velocita" network, formed by a north-south route from Milan-Rome-Naples, and by an east-west route from Torino-Milan-Venice. New tilt-body Pendolino (ETR 460) and non-tilting ETR 500 equipment with 300 km/h capability are being introduced to revenue service.

Spain

Spain has decided to adopt standard gauge for all new high speed lines to facilitate integration with a European network. TGV technology was selected for the new AVE Madrid-Cordoba-Seville high speed train, which entered revenue service in 1992. Near-term plans include the completion of a line from Madrid-Barcelona with extension to the French border. Longer term plans include 1,750 km of new or upgraded line, much of which will be served by Talgo Pendular equipment with 200 km/h capability.

Sweden

The X2000 tilt-body train entered revenue service on the upgraded Stockholm-Gothenborg route in 1991. The scheduled time for the 456 km trip is now 2 hours and 55 minutes. X2000 service has now been extended north to Sundsvall and south to Malmo. Near term plans include a Malmo-Gothenborg-Oslo-Karlstad link to form a network of high speed operations in southern Sweden.

X2000 technology has been tested and demonstrated both in Germany (to 275 km/h) and in the U.S. North East corridor, and is being promoted for implementation as a cost effective solution for HSGT in many intercity corridors worldwide. Demonstration runs have also been completed in Australia.

U.K.

British Rail is upgrading and completing the electrification of three main trunk routes; London-Birmingham-Manchester-Glasgow, London-Newcastle-Edinburgh, and London-Bristol-Cardiff. New IC 225 trainsets hauled by Electra powercars with 225 km/h capability are being introduced. The link from London to the Channel Tunnel, presently limiting Eurostar trains to 110 km/h, is under review.

1.5.2 Asia

Japan

For many, the Japanese bullet train represents the quintessential high speed train. Designed and built in the 1950's by means of a World Bank loan to re-establish the infrastructure of Japan and opened for revenue service between Tokyo and Osaka in 1964, the Shinkansen has carried over 1 million passengers in one day (in 1975) and almost 3 billion passengers over 30 years without a single passenger fatality (perhaps fortuitously in consideration of the infrastructure damage caused by the Hanshin earthquake of January 1995). The current Shinkansen service comprises the following lines:

Tokyo-Osaka (1964)-Okayama (1972)-Hakata (1975) ; 1070 km

Tokyo-Sendai-Morioka (1982) ; 496 km

Tokyo-Niigata (1982) ; 301 km

The Shinkansen continues to evolve. New Series 300 rolling stock has been introduced. A new line is being built to Nagano in time for the 1998 Winter Olympics. R & D projects include the development of 350 km/h Shinkansen equipment, and various advanced trains are being tested (including STAR 21, WIN 350 and 300X). Future plans include an extension of the Tohoku line to Hachinohe and, perhaps, to Sapporo, a new line from Tokyo-Toyama-Osaka (for which the Nagano Shinkansen is the first link), and an expansion of the Sanyo line from Hakata to Nagasaki and to Kagoshima in Kyushu.

The Japanese Maglev program continues apace. The first 18.4 km of a new 42.8 km test line is under construction in Yamanashi prefecture, west of Tokyo, at a cost of 300 billion yen (approx. U.S.$3 billion) for a 5-year development program. The double guideway line will be 70% in tunnel and will allow test vehicles to pass at high speeds. Detailed plans have been prepared for a new 500 km/h Chuo line from Tokyo-Nagoya-Osaka for revenue service in 2007. This will relieve congestion on the (then) 40 year old Tokaido Shinkansen line and will cut minimum travel time from 2.5 hours by Shinkansen to about 75 minutes by Maglev.

Pacific Rim

The industrializing countries of the Pacific rim represent a huge market for new transportation infrastructure. In the near-term, both Korea and Taiwan are planning high speed rail lines.

The Korean government is constructing a new line with 250 km/h operating speed from Seoul-Pusan, a distance of 400 km, for completion in 1998. TGV technology has been selected for implementation, and civil works are well underway. In Taiwan, planning studies have been completed for a 345 km, 250–300 km/h link between Taipei and Kaohsiung perhaps by 2000.

The economic development of China is presently inhibited by inadequate transportation services. The Beijing-Shanghai corridor is being examined for implementation of HSGT.

1.6. LOOKING AHEAD

Why the difference between North America and the rest of the world? American travelers to Europe and Japan are delighted with the fast, frequent, clean, comfortable intercity services provided by high speed rail. They return wondering—why not in the U.S.?

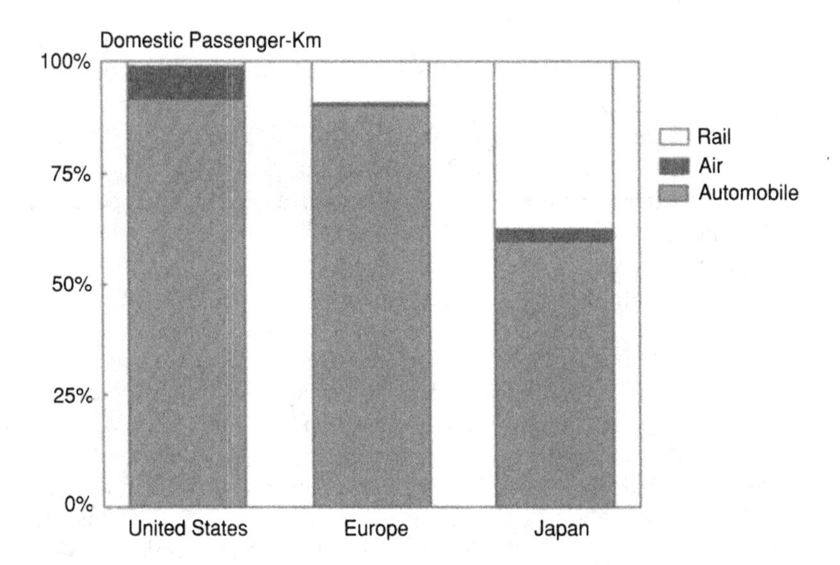

Figure 1.7 Modal distribution of domestic passenger travel in the United States, Europe and Japan.

Fig. 1.7 shows the relative modal usage for domestic intercity travel in the U.S., Europe and Japan. Automobile travel is clearly the dominant mode in all developed countries. An almost negligible number of passengers travel by rail in the U.S. (except in the North East corridor). This is perhaps both a consequence and a cause. The traveling public does not use rail because it finds the service nonexistent or inadequate, and much prefers the privacy and convenience offered by the automobile or the speed of air travel. Hard-nosed financiers are extremely reluctant to consider investment in new rail infrastructure, even in partnership with state and federal governments because of uncertainties in ridership and therefore in revenue and a perceived marginal return-on-investment.

North America has invested massively in road and air infrastructure over the last 40-50 years following the second world war. These modes have served the continent well, providing the facilities for inexpensive, convenient mobility of people, food products and industrial goods. However, mobility is being threatened by congestion (gridlock) on highways in major metropolitan areas and by delays and the limited available take-off and landing slots at hub airports (winglock). And the situation can only deteriorate into the 21st century as the economy continues to grow, albeit slowly.

Historically, rail transportation was used as a tool for regional economic development. It was the railroad that led to the development of the Florida coast and opened up the American mid-west. Today, the economic growth of the continent is being threatened by inadequate transportation. And yet, high speed rail is continually put to the test of economic viability (benefit/cost ratio and adequate return-on-investment, etc.). Governments insist that high speed rail must pay for itself, while dismissing the fact that the road and air modes have had and continue to receive massive subsidies through government allocations and taxation revenues. HSGT should not be regarded solely as a competitive mode, but as a partner to maintain mobility.

So, where is HSGT going in the next 20–30 years?

On the technology side, there is likely to be a convergence towards high speed train design with distributed motive power, integrated power electronics, and increased use of lightweight advanced materials for carbody structure to reduce dynamic loading of the track. Aerodynamic styling will continue to improve to slow the growth in motive power requirements as operating speeds increase. We are likely to see more and more cooperation between equipment and component manufacturers towards technical harmonization and the production of "world trains"—analogous to the global cars now being produced by some automobile companies. A reduction in the number of country-specific technologies, particularly in Europe, will reduce capital costs and enhance interoperability.

It is recognized that a primary factor inhibiting speed at ground level is aerodynamics. The power to overcome aerodynamic drag increases as the cube of speed. Noise from aerodynamic sources increases approximately as the sixth power of speed. And dynamic perturbations from aerodynamic transients such as passing and tunnel entry/exit become increasingly severe at high speed. It has therefore been suggested that high speed Maglev might benefit from being run in a tube. And if so, such a tube could be partially evacuated to decrease thrust requirements and energy consumption from aerodynamic drag. It has been further suggested that a transcontinental tunnel could link New York and Los Angeles, and onward under the Atlantic and Pacific oceans to provide the ultimate global transportation capability. Such a concept might even include a dipping and rising profile between stations to provide gravity assisted propulsion and braking for speeds up to 2000 km/h. Such ideas must however be dismissed as fanciful rather than visionary in view of practical engineering considerations such as the cost of construction and the maintenance of tunnel integrity. However, a 500 km/h system in which alignment curvature restrictions necessitate extensive tunneling might benefit from being operated in a low pressure environment in a continuous tube. Such a concept would seem to merit detailed techno-economic evaluation.

On the operational side, we can expect the steady enhancement of high speed rail systems in terms of speed, comfort and passenger services/amenities. European high speed rail services will become increasingly networked. More lines will be built in Asia, including the completion of a national network in Japan and new lines in Korea, Taiwan and China. North America is clearly a follower rather than a leader in the new high speed rail infrastructure, but the U.S. could see a rejuvenated passenger rail system, starting with the North-East corridor. However, there needs to be a greater threat to mobility to achieve this, such as steadily increasing road and air congestion as a limiting factor to economic growth or perhaps another oil crisis.

A mitigating factor may be the increasing use of telecommunications for video conferencing, thus reducing the necessity to travel and saving time and costs. However, there is no evidence yet that such communications facilities are slowing the growth of business travel. Notwithstanding impressive advances in telecommunications and the benefits which accrue from it, our cultures have not yet changed (evolved ?) to the stage at which the telephone, videophone, videoconferencing or, eventually, virtual reality are adequate substitutes for direct human interaction. So, HSGT must be a growth industry well into the 21st century.

REFERENCES

Community of European Railways (1989), "Proposals for a European High Speed Network".
Harrison, J. (1995), "High Speed Ground Transportation is Coming to America—Slowly", J. Transp. Engrg., 121 (2), 117–123.
Safford, M.A. (1990), "International Perspective on Intercity Passenger Transportation", Report No. SPA-90-1, Transportation Systems Centre, Cambridge, Mass.

TONY R. EASTHAM

Tony R. Eastham received the B.Sc. degree in Physics from the University of London in 1965, and the Ph.D. degree from the University of Surrey, England in 1969. After research positions at Plessey Telecommunications Ltd. and at the University of Warwick, he moved to Canada in 1972 where he joined the Canadian Institute of Guided Ground Transport.

Tony Eastham is now a Professor of Electrical and Computer Engineering at Queen's University in Kingston, Ontario, having joined the faculty in 1978. His research activities include transportation systems engineering, linear electric drives, and field analysis and intelligent design of machines and electromagnetic devices. Early research on superconductive magnet suspension and propulsion for intercity transportation resulted in the technical definition and evaluation of high speed systems for Canadian and U.S. applications. Linear induction motor (LIM) and linear synchronous motor (LSM) drives for urban transit and intermediate speed applications were then investigated both analytically and experimentally. Recent projects have included the evaluation of high speed rail and magnetic suspension systems, linear and industrial drives, non-contact energy transfer to highway vehicles, design and control of switched reluctance drives and permanent magnet synchronous machines, parameter identification in machines, and the development of CAD methods for machines and electromagnetic devices using the finite element technique.

Tony Eastham has conducted contract research and has been a consultant to a number of transportation organizations. Recently, he served as the only non-U.S. member on two U.S. Transportation Research Board Committees for the "Study of High Speed Surface Transportation in the U.S." and for the "Critique of the Federal Research Program on Magnetic Levitation Systems".

Tony Eastham is a Registered Professional Engineer in the Province of Ontario, a senior member of the IEEE and past Director of IEEE-Canada. He is currently serving as President-Elect of the Engineering Institute of Canada. He is Past-President of the Canadian Society of Electrical and Computer

Engineering, a Past-President of the Canadian Association of University Research Administrators, and a Past-President of the High Speed Rail Association. Administratively, he has served as Associate Dean in the School of Graduate Studies and Research and as Director of Research Services and Director of International Programs at Queen's University.

Upon completion of ten years in Research Services and International Programs at Queen's University, Tony Eastham was granted leave for one year. He spent the first six months (May-November, 1995) as JR Central Visiting Professor in Transportation Systems Engineering at the University of Tokyo in Japan. After a short lecture tour of South African universities, he began a six month engagement with the University Research Policy Directorate at Industry Canada in Ottawa in January, 1996.

2 CORRIDOR AND LAND-USE PLANNING CONSIDERATIONS

John A. Harrison, P.E.
Vice President
Parsons Brinckerhoff Quade & Douglas, Inc.
120 Boylston Street, Boston, Mass., 02116

William Gimpel, AICP
Senior Planner
Parsons Brinckerhoff Quade & Douglas, Inc.
250 West 34th Street, New York, NY 10119

2.1. INTRODUCTION

While land use and transportation are inextricably related, the nature of that relationship for any transportation mode is highly dependent on many factors, including a host of public policies. This Chapter will examine the most important of these factors and provide guidance on how to anticipate and resolve these issues when planning for deployment of a HSGT system.

2.1.1. The US Planning Framework

Historical Context

In the United States, there is a long tradition of separate constituent policies involving these two issues: land use has historically been regulated by states, primarily through their municipal jurisdictions, and transportation decisions have primarily been made through federal policies, especially its regulation of interstate commerce and the funding/subsidization of selected transportation modes, notably interstate highways, waterways and civil aviation.

Over the past 50 years, the expansion of "suburbia" has changed the focus of transportation policy and the distribution of transportation moneys from an urban to a metropolitan orientation, including the development of suburban/exurban traffic corridors reflecting the changes in locus of political power and influence. As a result, a lively public discussion about how to manage transportation investments to better predispose "public transportation friendly" land use development has taken place.

The decisionmaking process is the most salient difference between US and foreign transportation planning. National policy has dictated large federal commitments and support to high-speed rail system development in Japan and Western Europe. Within the US, where spending decisions have primarily been driven by the modal funding mechanisms in place (e.g., Highway and Aviation Trust Funds), HSGT development has yet to capture wide public attention, sufficient public subscription, and therefore adequate political consideration to assume a significant place alongside other favored transportation options in keeping with its potential as perceived by rail advocates.

While some of the single-ground-mode focus has been dispelled by the 1990 Clean Air Act and 1991 ISTEA legislation, the primary focus of this somewhat more inclusive "national transportation policy" has been metropolitan areas only (as a result of concerns vis-à-vis issues of non-attainment and intermodalism). The HSGT modal option will never prosper in the context of metropolitan planning. HSGT competitive strengths result from certain regional travel (for example, urban area-airport travel demand) or inter-city travel needs. Within North America there are probably at least a half-dozen or so attractive HSGT markets which have not been developed for lack of government sponsorship and a viable funding mechanism.

The lesson of history is that public policies and government regulation can either support or detract from the potential application of an emerging technology. Where promoted or permitted, new transportation modes and technologies have been effective in capturing markets. As an historical example, consider the replacement of streetcars by automobiles; and the replacement of trans-Atlantic passenger ships by air service. A current example of this phenomenon is Amtrak's significant market share within the New York-Washington corridor. It is doubtful that Amtrak could have captured its current market share without the active government support it has enjoyed.

The Transportation Planning Process: Politics, Economics, Technology

A variety of political and economic forces can combine to initiate the development of a transportation mode, including both infrastructure investment and technological innovation. When commercial land development interests and their supporters seek improved access to a city or regional core, that demand for transportation initiates a planning process which ideally results in an improved transportation system. This cycle has recurred frequently throughout the history of transportation development.

Consider the development of streetcar suburbs and beltway office parks. Transportation development is driven by the search to satisfy commercial demands for access to human and capital resources, access to consumer markets as well as facilitate exchange of market goods and services. Conceptually, we have an irrepressible human need to interact. Economic trade and the contact it requires can be seen as the dominant force promoting improved transportation technology.

The rail phenomenon may serve as an example of this progression. Much of the economic history of Europe and North America's former industrial might (and potential future might) is in the development of its rail technology. The United States had an undeveloped frontier accessed through one dominant passenger transportation technology/mode. HSGT is one of the most efficient and environmentally benign transportation technologies available to satisfy regional market demands if sufficiently supportive economic and political forces (public policy) can be mobilized to harness its potential in this country.

The political evolution of HSGT transportation options requires a perception about its role in catalyzing greater regional commercial interaction. Some of the best examples of HSGT successes are represented by "region-forging perceptions" with the resultant public policy development. For example, the perception that all major Japanese markets are within a day's journey of their major cities; the perception that connecting the three important cities in Texas, Ohio, and Pennsylvania would catalyze development; and the perception that building a radial transportation network around Paris, the self-proclaimed heart of Europe, would vastly improve trade and mobility. As long as all important stake-holders perceive that HSGT benefits will exceed HSGT costs, public transportation policy options evolve.

The globally dominant system of market capitalism derives from one simple rule: "who controls the most important capital investment rules". National or regional economies are about providing access to information and other "capital", including an educated work force and skilled labor. The historical industrial economy needed transportation links to fixed resources and large labor forces. Today's evolving service economy needs improved, more flexible access to dynamic, demanding and more specialized work forces and concomitant value-enhancing locations for their interaction.

The history of HSGT may be traced through several stages, perhaps most effectively through the development of high-speed rail in the late nineteenth century by rival railroads vying for premier intercity passenger trade. However, modern high-speed rail was invented by the Japanese to showcase their political, economic, and technical evolution at the 1964 Osaka Olympics. (This was paralleled by the Budapest premier of subway technology in Europe to showcase at a trade fair.) French TGV efforts paralleled and advanced from Japanese efforts to make the development of this technology a global phenomenon by 1972. Today HSGT development remains an effort to answer market needs through political-economic cooperation and support.

Essentially, currently successful HSGT efforts around the world represent this basic recipe of economic, political and technical cooperation. Distinguishing characterizations may be traced to subtle (and not so subtle) differences in geography, socioeconomic history, and leadership within the realms of economics, politics (policy) and technology. Unlike politics and economics, which are impacted by powerful symbolic forces like shared history, concept of nationhood, and desire for power, technology has a tendency to more uniformly develop across geographic and ideological divides. HSGT technology development is integrally related to certain economic and political forces, such as the level to which supported and related industries (e.g., rail car manufacture or steel production) constitutes the particular economy.

Nation building remains a powerful, albeit irrational, interest. Many infrastructure achievements throughout history (Roman roads, British navy, Russian and U.S. transcontinental railroads), attest to each generation's desire to apply a technological glue to bind a geography with a political psychology. The HSGT examples in Japan, France (alone and as a surrogate for Greater Europe), and Germany demonstrate this phenomenon.

Japan had a need for alternative access to domestic markets saturated by air travelers, and for re-investment in its industrial base. The Shinkansen was and remains a viable answer to that need. France's desire to support an established and vital rail industry corridor, saturated to the south of Paris, played a key role in the TGV's development. Germany's need to relieve airport congestion and to maximize rail infrastructure investment for passengers and goods contributed to the development of the ICE service. A constant attendant in this process is the need to avoid environmentally destructive infrastructure. Perceived American interest in relieving airport congestion, on oil import dependency, and to reduce the polluting consequences of an automobile-dominated transportation system may potentially play a crucial role in the the future development of HSGT options in North American markets.

2.1.2. Comparison with the European/Asian Planning Processes

France

In contrast with US transportation political/economic/technology dynamics, French national policies were motivated by perceived economic opportunities to develop an integrated transportation system, shaped and directed through conscious public policy. Relative to North American standards, this is a much more formal process. Planning is shaped, in part, by officially sanctioned government philosophies about land use development, resource allocation, and how the state should direct economic growth. The development of the TGV was a response to a corridor saturation problem. So many people want to travel between Paris and the Mediterranean that existing technologies and capacities could not satisfy that demand. Development of a larger network connecting European capitals and drawing a spider- web shaped network around Paris is less driven by market forces, and more the result of political/strategic considerations.

This process is directly linked to the French tradition for a strong central government with substantial control over domestic and international air travel, and with the will to manipulate market forces (domestic air travel prices) to direct travel. This is a clear example of national government policies used to support and promote a specific transportation policy. Important factors were the economy of land consumption, minimization of environmental (including noise) disruption, independence from oil imports, and promotion of key French industries and interests (rail and steel), and Paris' relative importance in Europe.

France's TGV is like other major improvements in transportation infrastructure: Its primary effects on travel, land values, and energy consumption are anticipated and often predictable, but secondary effects are dependent on the overall economic environment and government policies, as well as to how national firms and individuals perceive and utilize the new service.

Germany

Progression of events in last ten years have made for an interesting story in the development of HSGT in Germany. Reunification of Germany changed many factors very soon after the commercial premier of the ICE service. Post-war agendas included rebuilding the rail infrastructure including HSGT, but this policy was directed, in part, by the adjustments made necessary by Germany's partition. Much of the rail network in Germany had been designed in the nineteenth century with Berlin as the hub and with primary emphasis on east-west routes.

A further design criterion was to accommodate various operations over shared rights-of-way. This distinguishes Germany from the other examples. Emphasis on exclusive rights-of-way had been a criterion of the Shinkansen and the TGV networks from inception. Germany's ICE was designed to allow freight traffic to use the high-speed trackage at lower speeds. This flexible technology is related in principle to the tilt-train technology promoted in Sweden and Spain and has a mixed track record (the relative failure of Canada's LRC service using unproved tilt technology, for example).

Germany and Japan are set apart from other countries using HSGT technology in their continued strong support for the development of commercially viable magnetic levitation (maglev) technology. Unlike high-speed rail, high-speed maglev trains have not yet been applied on a large commercial scale, and continue to be criticized as a "technological solution in search of an economic problem." The critics may be proven wrong, however, if the currently planned Hamburg to Berlin maglev line is successful.

Japan

Following the near-total devastation of the Japanese industrial and economic base during World War II, a massive process of reconstruction induced the development of an improved national rail system. With a population of 125 million people in an area of 37 million hectares, Japan has an average population density of more than three persons per hectare. The 360-kilometer corridor between

Tokyo and Osaka has become the industrial and socioeconomic nucleus of modern Japan where almost half the population and two-thirds of the nation's industrial base is concentrated on only 16 percent of the total land area. Corridor congestion and network saturation have become the greatest transportation planning challenges of the post-war era for intercity carriers, including the railroads and airlines, and this problem has resulted in creative transportation planning solutions.

The debut of the Shinkansen in 1964 virtually revived Japanese rail travel, and redefined the term "high-speed" for the industry. The systematic research and development of a high-speed rail technology had been a conscious post-war national transportation policy objective and by 1958 a national committee had decided that a dedicated double-track high-speed rail connection would be the most effective solution to the Tokyo-Osaka Tokaido corridor's saturated passenger transportation capacity. The Shinkansen was enthusiastically received by the public and some trains were at full capacity within months. By 1979, the Shinkansen was earning one-third of the total revenue of the national railroad, and the original trunk line had been expanded three times.

The Shinkansen high-speed trains were truly revolutionary when introduced in 1964, and they have remained a commercial success (JNR's onerous debt problem notwithstanding) but lingering concerns about environmental issues (especially noise) have never been completely overcome. As in France, national policy in Japan has promoted the manipulation of air travel market values to shift demand from crowded Haneda airport. Also, a conscious effort to bind all distant parts of Japan into a day-trip network, was always part of the Shinkansen's justification. Ease of access to Osaka for the showcase 1964 Olympics served, in part, to show that the former aggressor nation had applied its technological and industrial leadership peacefully.

2.1.3. Summary of recent US Hsgt studies

High Speed Ground Transportation has been the subject of numerous feasibility studies over the past decade in various corridors around the United States. A summary of the status of many of these studies is provided in Table 2.1.

TABLE 2-1. Summary of Recent HSGT Studies in the United States.

Corridor	*Length km(mi)*	*Current Amtrak Servicefrequency*	*Status of high-speed rail studies*
Boston-New York City-Washington. D.C. (Northeast Corridor)	733 (455)	Boston-New York: 10 trains each way daily. New York-Washington, D.C.: 36 trains each way daily, including 15 Metroliners	The NEC has been the focus of U.S. HSR passenger rail improvements since 1976. with over $2.5 billion spent on rehabilitation and upgradng to date. An additional $2.4 billion is programmed. Current emphasis is to complete the electrification from New Haven and Boston and eliminate traffic bottlenecks. With the advent of new HSR tilt-body equipment, trip times of 3 hr Boston-New York and 2.5 hr New York-Washington will be possible.
Washington, D.C.-Richmond-Raleigh-Charlotte	770(479)	Seven trains each way daily to Richmond; two to Charlotte	Recently designated a section 1010 HSR corridor; eligible to receive federal funding. Current maxmum speed: 130 km/h (79 mi/hr); proposed maximum: 150 km/h (90 mi/hr).
New York City-Albany-Buffalo-Niagara Falls	744(462)	Eleven trains each way daily to Albany, including four through to Buffalo	New York State has invested heavily in its Empire Corridor service. The New York City-Albany portion is operated at speeds up to 180 km/hr (110 mi/hr), qualifying it to apply for federal HSR assistance under ISTEA. The New York State Energy Research and Development Auhority (NYSERDA) recently completed a high-speed-surface-transportation feasibility study.

Corridor	*Length km(mi)*	*Current Amtrak Servicefrequency*	*Status of high-speed rail studies*
Dallas-Ft. Worth-Houston-San Antonio	Dallas-Houston: 402 (235) Houston-San Antonio 345 (210)San Antonio-Ft. Worth: 467 (2X5)	Daily train service via long-distance trains.	Texas High-Speed Rail Authority awarded an exclusive franchise to Texas TGV to design, build, and operate a state-of-the-art HSR system in the Texas Triangle without the use of state funds. Texas TGV defaulted and the franchise offer was rescinded.
Los Angeles-San Diego (Lossan Corridor)	210(128)	Sixteen trains per day each way	Outside of the NEC, highest passenger train frequency of any Amtrak route. Portion of the corridor currently operated at 150km/h(90mi/hr).
Los Angeles-Bakersfield-Bay Area/Sacremento	680 (415) to San Francisco; 640 (390) to Sacramento	Four trains each way daily from Bay Area to Bakersfield; one train daily along the coast between San Francisco and Los Angeles. Seven trains each way daily between Sacramento and a the Bay Area.	A 1996 HSR Summary Report and Action Plan by the California High-speed Rail Commission determined HSR to be feasible
Eugene-Portland-Seattle-Vancouver B.C	760(464)	Four trains daily each way including two TALGO trains. (Two trains daily each way between Vancouver, B.C. and Eugene did one each way between Vancouver, B.C. and Seattle and between Seattle and Eugene.)	The Pacific Northwest Corridor was designated as a section 1010 corridor. The states of Washington and Oregon are proposing a program of *incremental* improvements to raise passenger train speeds and frequencies. The first phase will raise top speeds from 130 km/h (79 mi/hr) to 150 km/ h (90 mi/hr). A second phase would raise the top speed to 200 km/h (125 mi/hr). Three TALGO tilt trains have been ordered to expand this popular service

Philadelphia-Harrisburg-Pittsburgh	565(351)	Eight trains each way daily to Harrisburg, including two through to Pittsurgh .	Pennsylvania HSR Feasibility Study contracted in the mid-1980s considered conventional HSR and maglev. Current proposals by Pittsburgh-based group vying for federally funded maglev demonstration.
Designated section 1010 HSR Corridors			
Chicago Hub Corridors Chicago-Detroit	449(279)	Four trains each way daily to Detroit; two trains daily to Kalamazoo.	Detroit-Chicago Rail Passenger Corridor Develpmental Blueprint, completed in 1991. recmmended a phased program with maximum running speeds of up to 200 km/h (125 mi/hr).
Chicago-Milwaukee-Twin Cities	700(435)	Eight trains each way daily to Milwaukee; one train daily to Minneapolis/St. Paul.	Tri-State HSR Study completed in 1991 recomended 200 km/h (125 mi/hr) Amtrak upgrade. Maglev also being considered. Section 1010 proosal for Chicago-Milwaukee portion (140 km, 85 mi) is to achieve 150 km/h (90 mi/hr) pasenger train speeds.
Chicago-St. Louis	454(282)	Four to six trains per day each way.	State of Illinois has studied upgrade of passenger service to 220 km/h (135 mi/hr) with average speeds of 135 km/h (83 mi/hr). Corridor contains 325 grade crossings. Preliminary plan calls for closing 231 crossings, constructing 39 grade separations and upgrading the protection at 56 loations.
Tampa-Orlando-Miami	411(255)	Two trains daily each way, plus connecting bus service.	Franchise proposals of the late-1980s failed. State of Florida still interested in pursuing HSR and has selected a private consortium led by GEC Alsthom and Bombardier to negotiable a franchisee to plan, design, build and operate a system. Designated as a section 1010 corridor.

2.2. POLITICAL AND ECONOMIC CONSIDERATIONS

2.2.1. Market-Driven Process

Endpoints Define Route

The fundamental issues that determine HSGT routes are where people are and where they want to go. These endpoints define the whole exercise of transportation planning and development. Intermediary points of varying importance and influence will vie for alternative routes or stations, but an HSGT route cannot compromise its opportunities for commercial success by ignoring the big market magnets such as downtowns or airports which invariably tend to define its endpoints.

There is ample material for discussion about central place theories over time, and the long-term demand for access to central cities in different cultures. Transportation planners can be fairly confident in the long-term market attractiveness of central Tokyo and Paris, but the nature of North American "metropolitanism" and the prevailing cultural values about space and interaction across the economic class lines of urban/suburban areas leaves open the transportation issue of America's future downtown or metropolitan market access.

Some people argue that the important interactions between US "movers and shakers", who traditionally rely on person-to-person hand shakes and eye contact to make market decisions of great consequence, will always take place in the traditional downtowns. Other assert that telecommunications and suburban centers will surpass the traditional pull of the US downtown. Considering the potential long-term attractiveness of Washington's Tyson's Corner or Houston's Oak Park, for example, that argument may have considerable locational merit for some areas.

Depending on HSGT technology, speed, and topography, optimal/acceptable passenger service over given distances will be measured by time in arrival. Therefore, city endpoints practically suitable for HSGT linkages may be selected by their relative distance. HSGT is not practical for New York-Seattle, but promising for Chicago-St. Louis, or Los Angeles-San Francisco.

Non-commercial destinations could be included in HSGT plans if sufficient demand develops for such travel. That demand may be a function of dominant cultural attitudes about travel behavior (habits of travel). Most Americans do not consider taking a train for a ski trip or seaside holiday, but in Japan or France vacation packages to such locations are well supported by travel habits, and contribute to the commercial success of the relevant HSGT systems. In such cases, the attractiveness of one resort or facility over others may become an open rivalry, beneficial to the traveling public. However, snow-capped mountains and ocean beaches are fixed features on the market map, and can obviously serve to define route endpoints.

Station Location and Spacing Shape Route

Station locations may be determined by several factors, including politics, economics, and technical adaptability. Perhaps the most important factor is the travel demand for endpoint markets, and traveler's willingness to compromise their travel time objectives (between the major markets) to accommodate the intermediate transportation needs of short-term travelers. For that reason, station spacing may be shaped, in part, by HSGT technology's ability to recover from these dwell time delays by more rapid acceleration from stops. If there is sufficient travel demand for several intermediate stops on a route, service options may have to be built into route operations to allow sufficient flexibility for several operational patterns to function simultaneously, including passing sidings, off-line stations, and other design features.

2.2.2. Interaction between Political and Economic Force

Impacts of Transportation Improvements on Land Values

Perhaps the most significant and pervasive institutional factor which requires consideration is the issue of land. It becomes the most significant issue when viewed from the perspective of availability of land for HSGT systems requiring new rights-of-way. At the same time, the effects which HSGT system implementation could have on current and future land use patterns also affect the social and political acceptability of the systems. To deal adequately with land as a factor in HSGT implementation, both of these aspects require detailed examination.

The taking of land for public transportation facilities has met with increased resistance by the public at large. There is reason to believe that land use in the future is likely to be the most serious institutional obstacle to the development of new transportation systems. Community opposition to airport expansion and new airport site selections have become commonplace news. Highway and freeway projects have been stopped by concerned citizen groups not only in urban areas but in sparsely populated areas as well.

A new HSGT system, with complementary adjustments in land- use regulations (generally allowing greater development densities and/or activity mixes), could mean substantial changes in land development patterns. Conditions that led to increased HSGT patronage and concomitant declines in automobile travel could potentially slow the sprawling development of some exurban areas over time. Assuming coordination between municipal and state land-use planners and federal and state transportation policy objectives, over time (no small assumption), US land development patterns could begin to resemble European/Asian patterns. In any event, such coordination would assist HSGT prospects.

The best examples in the United States of new development along intercity rail corridors might be Metropark, New Jersey and New Carrollton, Maryland. However, these sites are also well serviced by highway facilities. The limited development of HSGT in the US has seen, or resulted in, few examples of European/Asian HSGT land development.

Potential for Station-Area Land Development

Many factors potentially affect station-area development including station access and egress options (pedestrian, public transportation or parking facilities), station operational space requirements, joint development commercial opportunities, municipal predisposition to consider and rezone to more intense uses around stations, various station area development patterns emerge.

Washington Union Station is a mall with a food hall, a cinema, a train station, a metro station, a bus station, taxi stands, and a parking structure. Many of this facility's users benefit from the scale economies of such mixed use agglomeration without consideration of other travel mode options. Some stations, like Pennsylvania Station Newark, have become business destinations because of the critical mass of office space within easy walking distance of the train station (via a private sky way). Other stations, like New Carrollton, have become destinations due to development of substantial office space densities conveniently accessible to the station (again, through private connections).

Opportunities for Public-Private Partnerships

Private money may be involved in HSGT development under various circumstances:

- there is money to be made from private investments based on projections for demand that will amortize debt and pay returns within a reasonable amount of time (conventional capital investment);
- there is money to be made from the development of the land around the stations or from the operation of built facilities, such as retail or parking services (conventional public-private partnership); or
- there may be contracting opportunities or other potentially lucrative business opportunities in the facility's design or construction, opportunities that may be realized only after an initially unprofitable investment into the venture that secures a role in its operation (e.g., Japanese "Third Tier" investment arrangements). The Third Tier method is relatively new in Japan and has attracted considerable criticism. Such arrangements may not be politically acceptable in the United States.

Local Demand for Service

It is not clear whether a HSGT station alone can stimulate a local demand for the service. Saint Pierre des Corps, well outside Paris, remains a fancy HSGT station in a small, non- descript town. Most local travelers wait for the non-premium service unless they are making a special trip. This HSGT commuter phenomenon is not unheard of in France (particularly between Paris and Lyons or Marseilles) or Japan (especially between Osaka, Nagoya, and Tokyo), but is not very common at other intermediate/local points due to the high development costs and uncertain service prospects. Any sacrifice in local service to accommodate HSGT service is likely to breed local resentment and project resistance among prospective supporters.

Effects of Alignment on Surrounding Land Uses

a. Relative Costs and Benefits of Alignment Construction

HSGT systems utilize at-grade, elevated and tunnel construction depending on the surrounding terrain and land use. Each type of construction has inherent advantages, disadvantages and capital and operating and maintenance costs associated with it. At the risk of over-simplification, elevated construction for high-speed rail is generally roughly twice as expensive as at-grade construction and about one-half to one-third the cost of tunnel construction. For maglev systems, the cost of at-grade construction is generally only slightly less expensive than elevated guideway and tunnel construction is two to three times more costly. The relative advantages and disadvantages of each are summarized in Table 2.2.

TABLE 2.2. Relative Advantages/Disadvantages of HSGT At- Grade, Elevated and Tunnel Construction Types

	Disadvantages	*Advantages*
At-Grade	potential for land parcel/ community isolation/restricted access; increased number of grade separations required; and potential for significant construction impacts	lower construction costs; fewer speed consequences of changes in grade; easier access to platform by passengers; and smaller visual envelope.
Elevated	greater construction costs; noise amplification; larger visual envelope; and potential for objection to visual impacts of structure, especially shadow below.	increased safety; less divisive influence on local land pattern; small footprint; and reduced vulnerability of right-of-way to intrusion and consequent delays.
Tunnel	greatest construction costs; difficult emergency egress, and tendency to complicate station arrangement and access/egress from platforms.	increased safety, except emergency egress; less divisive influence on local land pattern; reduced vulnerability of right-of-way to intrusion and consequent delays; and reduced noise consequences.

b. Environmental Considerations

As with any major transportation capital investment, there are many route alignment environmental issues to be considered. These issues include:

- air quality issues, including automobile pollution abatement, and power plant emissions (supplying power to HSGT trains);
- noise and vibration issues. These issues have been particularly important in the development of Japanese and French HSGT systems;
- electromagnetic fields (EMF) and other energy field issues, applicable to high-speed rail as well as maglev technologies;

- energy consumption issues, including reduced automobile petroleum demand, and HSGT energy requirements;
- wetland protection issues, and other environmentally sensitive natural resource concerns;
- displacement and financial issues, and any other land-use concerns, including relative distribution of costs among economic classes, and among users and non-users;
- visual impacts; and
- construction impacts.

Environmental Review/Approval Considerations

A primary consideration in planning any new transportation project is its effect on the environment. While railroads in general, and HSGT in particular, are environmentally less polluting and consume less land than highway or air modes, there are drawbacks, primarily noise and vibration and EMF, associated with this mode that raise concerns and/or outright opposition to building or up-grading high-speed passenger lines. The NEPA public involvement process and the courts allow project detractors to voice their opposition, often delaying and/or killing otherwise viable and worthwhile projects. Planners must carefully choose routes that will absolutely minimize environmental impacts to the natural and human environment and mitigate any unavoidable impacts wherever possible. This adds to the cost and environmental review/approval time, to the detriment of many projects. The best recourse is to thoroughly investigate and document alternatives to the proposed siting or design approach, being able to defend the proposed design as the best that can be achieved. Even that may not be enough, but at least it greatly improves the chances for project success.

a. NIMBY Forces

Not in My Back Yard (NIMBY) forces must be contended with in planning for HSGT and, in fact, are common to any transportation development effort. Because HSGT would provide premium service to wealthier citizens and would not be perceived as a benefit to a poor community, any effort to run the line through poor neighborhoods will likely stir concerns about "environmental justice". This kind of debate is often clouded with many related, but tangential issues, like class equity, recollections of past battles (veterans of 1960s highway debates), and strong distrust of the evaluation methods used to make decisions due to their perceived biases.

HSGT may encounter stronger NIMBY resistance than other transportation modes because the public is generally unfamiliar with this technology, is concerned about noise and vibration, and has persistent fears about EMF.

2.3. TECHNICAL CONSIDERATIONS

2.3.1. Where does HSGT Make sense?

Planning for high-speed ground transportation (HSGT) starts with a need to transport thousands of people a day in an inter-city corridor a distance of between one hundred and four hundred miles long. That is the range where HSGT is most trip-time and cost competitive with auto and air. Shorter corridors do not need high speed and can be well served by conventional rail; longer corridors cannot compete effectively with air. Lesser volumes of passengers do not justify the large investment in infrastructure and equipment needed for implementing HSGT.

Corridor Planning Process

HSGT corridor planning is by necessity a long-term process involving a myriad of stakeholders. Typically all levels of government need to be involved, as well as the private sector. Such large-scale undertakings require years of planning, design and construction before the first passenger can be carried. Discussed below are HSGT route planning considerations.

Planning for high-speed ground transportation systems, as for any fixed guideway mode, involves identifying feasible routes to connect desired station sites. The challenge of routing HSGT systems is to find suitable corridors that are straight and flat enough to accommodate high speeds. Often existing transportation (highway or rail) rights-of-way are proposed for HSGT to minimize the environmental impact and cost of assembling the required land. But the geometric characteristics of existing routes are normally inadequate to support high speed, and the cost of grade- separating and reconstructing existing highways to accommodate an HSGT system running adjacent to or in the median of a highway may offset any monetary advantage otherwise available. Nevertheless, considering the highly urbanized nature of major segments of the corridors that are being discussed for HSGT implementation in the U.S., extensive use of existing rights-of-way at least in urban areas is inevitable. Compliance with the National Environmental Policy Act (NEPA) and other federal and state laws and regulations requires any environmental impacts resulting from a project to be carefully considered and compared with alternative solutions and subjected to public review and comment before any proposed project is allowed to proceed.

Because no new HSGT system has progressed past the early preliminary engineering stage in the U.S., there is no precedent for or experience in siting new high-speed ground transportation system alignments in this country. One has to look to Japan or Europe for examples of actual experience in this domain. In numerous studies of proposed high-speed routes in various candidate corridors across the country, however, planners have been challenged by the difficulty and complexity of identifying feasible routes, and in the process have developed some new computerized techniques that will greatly improve the siting and route planning processes.

a. New GIS/CADD Techniques

These tools, that rely on Geographic Information System (GIS) data bases and Computer Aided Design and Drafting (CADD) techniques, provide a vastly improved capability over the former relatively primitive route planning techniques that relied solely on information contained in United States Geodetic Survey (USGS) paper maps (Quad Sheets). Paper maps can be used for the initial screening of potential routes, but they are typically never up- to-date and do not reveal sufficient land use/land cover information for reliable analysis. Up-to-date satellite imagery, USGS Digital Line Graphs (DLGs), Digital Elevation Models (DEMs), and other sources of electronic geographic data, such as Census Bureau TIGER data, provide a wealth of information that can be used effectively in identifying planning alignments for HSGT systems. Then, once suitable corridors are identified, more detailed mapping and alignment design can proceed with improved assurance that the routing selected will be feasible.

The corridor planning work undertaken in the spring of 1995 for the Federal Railroad Administration's Commercial Feasibility Study developed and employed these new GIS planning tools for identifying and characterizing feasible illustrative alignments.

2.3.2. Selecting the Right Technology

One of the most important factors affecting the success of any proposed high-speed corridor development project is the technology to be employed. A range of technologies are available from conventional rail to Very High Speed Rail to advanced technologies including high-speed magnetically levitated (Maglev) trains. For the Commercial Feasibility Study, nine technology options were considered and were evaluated for use in candidate high-travel-density corridors around the country. These technologies fell into the four basic categories described below:

- ***Pre-High Speed:*** Conventional passenger trains operated on existing freight railroad trackage with appropriate track improvements at speeds up to 79 mph and with proper signaling equipment at up to 90 mph.
- ***Accelerail:*** Incrementally improved passenger train service operated generally in the speed range from 110 mph to 125 mph and in some locations at up to 150 mph (track alignment permitting) with either all-electric or fossil-fuel- powered locomotives. As in the Pre-High Speed case, this option is based primarily on sharing freight railroad trackage or at least existing railroad rights-of-way.
- ***New High Speed Rail (HSR)*** (sometimes referred to as Very High Speed Rail): An entirely new state-of-the-art 180–200 mph rail system operated on its own fully grade-separated and dedicated tracks using high-powered electrified trainsets similar to the successful French TGV or German ICE technology. In urban areas new HSR could share rights-of-way with other railroads. In suburban and rural areas, this mode would likely follow interstate highways or existing railroad corridors, but on its own independent alignment that would have to be sufficiently straight and flat to permit 200 mph sustained operation.
- ***Maglev:*** An entirely new advanced-technology magnetically-levitated train capable of 300 mph cruising speed on its own fully grade-separated, dedicated guideway. As in the new HSR case, in urban areas, the Maglev guideway could share interstate highway or railroad rights-of-way and would be speed-limited in cities. In suburban and rural areas, this mode would likely follow interstate highways or have an independent alignment, but would be operated on its own generally-elevated guideway that would have to be sufficiently straight and flat to permit 300 mph sustained operation.

Track Geometry Limits Speed

Each of these four basic categories of technology has its own associated performance and track/guideway alignment criteria based primarily on ride-comfort considerations, which are generally more restrictive than safety-related speed restrictions. Current and historic U.S. federal regulations are quite conservative in limiting unbalanced lateral acceleration to about 0.05g. Vertical acceleration is not regulated, but railroads still follow historic rules that were originally established around car coupler limitations. Modern suspension and car coupling systems permit higher vehicle speeds in (vertical and horizontal) curves without creating passenger discomfort or endangering passenger safety.

To increase speed capability in curves, without sacrificing safety or comfort, vehicles can be designed to employ active tilt-body suspension systems as demonstrated successfully on the Italian "Pendolino" and more recently on the Swedish X2000 trainsets. While not currently employed in Very High Speed Rail (e.g., 200-mph systems) or in current 300 mph Maglev prototype designs, such systems would limit the lateral acceleration felt by passengers in curves by tilting the carbody, creating higher effective superelevation in curves, thus allowing higher speeds without discomfort. The federal Commercial Feasibility Study assumed "Tilt-train" technology would be used under eight of the nine cases studied – every one above the 79 mph base case.

Variations in vertical acceleration can cause motion sickness when the variations occur at the specific frequencies near 0.2 Hertz. [3-1] Steady state vertical acceleration on the order of 0.2 g would be acceptable, but variations in vertical acceleration, especially in the 0.2 Hertz frequency range, would result in an unacceptable "roller-coaster" ride. This is important in defining highway crossings or high-speed high-bank angle curving as is possible with Maglev. For repeated events occurring more frequently than every three minutes, the Commercial Feasibility Study assumed that the apparent vertical acceleration be limited to 0.1g as measured as a deviation from the normal gravity field. For frequent curves, this meant limiting total bank angle to 20 degrees except in very infrequent instances where up to a 30 degree bank angle was judged to be acceptable in an occasional large (long) curve.

In practice, the real limitation to introducing horizontal and/or vertical curvature in High Speed Ground Transportation system alignments is not steady-state vertical or horizontal acceleration comfort limits, but the rate of change of acceleration or jerk rate and the roll rate at which high bank angles can be achieved. Jerk rates of less than about 0.07 g/sec. and roll rates of less than 5 degrees per second are generally believed to be necessary to maintain passenger comfort, minimizing the possibility of passengers losing footing while standing or walking about the train. Higher acceleration and jerk rates could be achieved without great discomfort if passengers were constrained (strapped into their seats) with seat belts and shoulder harnesses, but such an aggressive seating policy/design approach is not currently planned or envisioned to be necessary.

New HSGT versus Incrementally Improved Existing Rail

The FRA's Commercial Feasibility Study considers two categories of alternative infrastructure and equipment improvement scenarios (Pre-High Speed and Accelerail) that are proposed to be developed in existing predominately freight railroad corridors. The travel time potential of these scenarios can be assessed by evaluating such existing railroad parameters as track alignment and condition, signal system type, number and location of grade crossings and so forth, and estimating the railroad system upgrade potential. To understand this approach to providing high-speed rail passenger rail travel, it is necessary to appreciate how "Accelerail" differs fundamentally from "New HSGT."

a. Existing Railway Alignment Realities

The presence of curvature, and to a lesser extent gradients, in an existing railroad alignment and profile poses a fundamental limitation on its maximum speed potential. Electrification, signaling, grade crossing elimination and sophisticated equipment cannot overcome the ride comfort and, in the extreme, safety limitations imposed by grades and curves (except for the slight advantage provided by tilt-bodied trains, and this only when the tilt system is properly functioning). Grades and curves arise in existing railroad track alignments for a variety of reasons including topographic constraints, principally related to terrain – the existence of natural barriers, such as mountains, rivers, lakes and wetlands – and urbanization, environmental, and other land-use constraints, which often dictate where railroad facilities can or cannot be located. Also, the state-of-the-art of railroad technology that existed when the railroad was built, i.e., construction techniques and type of motive power and rolling stock available, and the use for which the railway was intended, i.e., type of service to be provided, often dictate the allowable gradient and curvature that were deemed acceptable for each rail line. Most railroads are laid out with a target maximum speed for a given type of operation, which dictates allowable grades and curves. Cost of construction, service requirements and performance goals all affect how relatively strait and flat a railroad track alignment must be. It is often less costly to go around a natural or manmade impediment than to go over, under or through it. So, the combination of physical (including environmental) constraints, railway service needs, available technology and cost invariably drives the location and physical geometry of railway track construction.

Another constraint to high-speed operation in most railroad corridors is the existence of frequent at-grade highway crossings. The cost of fully grade separating or closing all crossings in any particular corridor is prohibitive, and in many cases it may also be infeasible. There is active research, demonstration and testing of improved grade crossing warning and protection systems currently on-going, but even with advanced warning systems, under no circumstances will crossings be permitted on routes operated above 125-mph. Therefore, the existence of grade crossings will continue to constrain rail line speeds in many Pre-High Speed and Accelerail corridors for many years in the future.

b. Speed in Curves

The laws of physics ultimately dictate what performance can be achieved on any given track alignment. One of the primary constraints to achievable track speed is curvature that introduces centrifugal forces and resultant lateral acceleration in proportion to the square of the speed and inversely proportional to the radius of curve. Lateral acceleration can be counteracted by track superelevation in curves, i.e. by raising the outer rail in a curve above the inner rail to counteract the centrifugal force. This has the similar effect as the banking of a highway curve or a race track turn. There are practical limits to superelevating railroad curves, however, particularly when considering mixed freight and passenger usage. Freight trains that traverse a curve at less than "balance" speed impose more weight on the lower rail causing increased rail wear, and in the extreme can create a potential unsafe condition – overturn or derailment – particularly with high-center-of-gravity cars, e.g., double stack, trailer on flat car (TOFC) and tri-level autorack cars, which are increasingly prevalent on U.S. railroads. Generally, freight railroads will not superelevate track in curves more than about 4 inches, even though regulations permit up to 6 inches for passenger trains in the U.S.

In addition to actual superelevation built into track, trains typically operate with some limited "unbalanced" elevation – i.e., the equivalent of the amount of additional superelevation that would have to be built into the track to reach a "balanced" equilibrium situation. Tilt trains compensate for high unbalanced superelevation by rotating or "tilting" about their center of gravity so that the lateral force felt by the passenger is equivalent to the force felt in a conventional non-tilting train at lower speeds. The result is that tilt trains can traverse a curve 30 to 40% faster than a non-tilt train at an equivalent comfort level (i.e. equivalent unresolved lateral acceleration).

It is important to understand the fundamental relationship between design speed and minimum curve radius. Allowable curve radius increases with the square of speed, requiring very large radii curves to support high speed. So, for example, for a train to go 200 mph, the minimum radius of curve, assuming an equilibrium elevation (sum of superelevation and unbalance) of 12 inches, would be 13,200 feet (which equates to a 0 degree 25.7 minute curve). For new high-speed alignments in rural areas with flat terrain, such large radii curves can generally be achieved. But in urbanized areas or where topography or use of existing rights of way constrain the alignment, high-speed alignments can be difficult if not impossible to achieve.

c. How "Accelerail" Differs from "New HSGT"

Accelerail routes are fundamentally constrained by the existing railroad track alignment with modest improvements in track structure, signal systems, and motive power. Generally each curve is analyzed for "straightening" potential, considering degree of curvature, length of curve and the presence of adjacent curves, as well as topographic, environmental and other physical features that constrain the alignment. In the Commercial Feasibility Study, it was assumed that track shifts of twenty feet or less could be accommodated within existing rights-of-way. Shifts of greater than twenty feet were evaluated on a case by case basis to see if they would be potentially feasible. Engineering judgment was used to assess the relative feasibility/unfeasibility of each major curve realignment, and only those judged to be feasible were included in the proposed upgraded railroad scenarios. Implementation of "Accelerail" improvements would require an in-depth and detailed assessment and accommodation of existing conditions.

New HSGT planning alignments, on the other hand, are based on achieving a large minimum radius curve at the desired design speed, wherever possible, generally at greater expense than in the Accelerail cases. Costs must be incurred to acquire and prepare new ROW, and construct earthworks, structures, track/guideway, signaling and electrification systems and so forth. Except in urban areas, and in a few special circumstances, new alignments can be planned for the desired maximum design speed. So while the estimated required costs are much higher for new HSGT alignments, the system performance is also considerably improved.

The two approaches are fundamentally very different – Accelerail cost and performance depend on the existing railroad conditions and alignment, whereas new HSGT performance is essentially pre-set and the costs are those associated with achieving this high performance. They are not on a continuum. The Accelerail scenarios represent the best that can be achieved by incrementally improving the existing railroad; new HSGT can be built to much higher standards, but at much higher cost.

2.3.3. Capital Cost

Capital costs for HSGT systems are by their nature corridor and technology specific. The major categories of cost for new HSGT systems are for land acquisition, earthwork, structures, systems, vehicles, and program implementation. Typically, each of these categories is broken down into subcategories, each made up of constituent cost elements. Unit costs are generally assigned to each cost element to prepare an overall estimate.

Capital costs – the invested cost for infrastructure and equipment – differ depending on the technology selected. Infrastructure costs give a good indication of system cost differences. Except for the Pre-HSGT and Accelerail cases, trainset costs are generally less than 10 percent of all costs and are considered to be about the same for the various high- speed ground technologies.

HSGT system costs increase as the design speed increases. At the low end, for incremental improvements to existing railroads, the primary costs are for track and signal upgrades, grade crossing treatment and new rolling stock. As speeds are increased above the level that can be supported by existing infrastructure, the cost of right-of-way acquisition, earthwork and structures soon become the dominant cost items. Earthwork and structures (bridges, viaducts and tunnels) generally account for more than half of new HSGT system costs.

Infrastructure costs for maglev are generally estimated to be about 30 to 40 percent higher than for very high-speed rail, primarily as a result of maglev's more expensive system elements (guideway, signals, communications and electrification).

TABLE 2.3. Conceptual Capital Cost Ranges (Million $/Mile) for HSGT Systems by Type of Construction.

TECHNOLOGY Speed Range	*HSR (TILT TRAIN) 125-150 mph*	*VHSR 150-200 mph*	*MAGLEV 200-300 mph*
At Grade	$6 - 10	$8 - 15	$20 - 30
Elevated	$20 - 25	$20 - 30	$25 - 40
Tunnel	$60 - 100	$60 - 120	$65 - 125

REFERENCES

Draft Memorandum dated February 17, 1995 from Michael Coltman, DTS-701 to Ron Mauri, DTS-401 (both of the Volpe National Transportation Systems Center, Cambridge, MA) entitled, "Summary of Technology Options for the Commercial Feasibility Study."

Los Angeles to Bakersfield High Speed Ground Transportation Preliminary Engineering Feasibility Study, "Alignment Criteria and Standards," prepared by Morrison Knudsen Corporation and Parsons Brinckerhoff for Caltrans District 7, December 1994.

JOHN A. HARRISON

John A. Harrison, a Vice President and Senior Professional Associate of Parsons Brinckerhoff, has 27 years of diversified engineering and management experience on railroad and transit systems projects, with expertise in systems engineering, operations planning, and project management. As Chairman of the Transportation Research Board's Committee on Guided Intercity Passenger Transportation, Mr. Harrison is regarded as one of the country's foremost authorities on high-speed rail and maglev technology. he has contributed to numerous high-speed rail studies at the federal and state level in the areas of technology assessment, cost estimation, safety and environment. He has also consulted internationally in Taiwan and South Korea on high-speed rail.

Prior to joining PB in 1983, Mr. Harrison served as head systems engineer on the Northeast Corridor Improvement Project, managed a year-long railroad electrification study for the Federal Railroad Administration, served as resident engineer of track construction on Washington Metro, and worked as a sales and applications engineer for General Electric Company's Transportation Systems Division.

Mr. Harrison holds a Bachelor's degree in Civil Engineering from Carnegie-Mellon University and a Master's Degree in Management of Technology from the Massachusetts Institute of Technology. He is a licensed professional engineer in Massachusetts and four other states.

WILLIAM GIMPER

William Gimpel specializes in transportation planning and comparative urbanization. He is experienced in regional and municipal planning and policy analysis, including work in program and project evaluation, transit planning, goods movement planning, and environmental documentation. Mr. Gimpel has focused on the effects of transportation investment upon economic development and urban form, with extensive investigation into relationship between transportation and land use

After graduating from the College of William & Mary in 1984, Mr. Gimpel served in the Peace Corps as a UNICEF volunteer and mathematics teacher in Ghana for two years. He later earned two master's degrees from Columbia University, one in International Affairs and one in Urban Planning. His master's thesis, concerning the potential land-use effects of high-speed rail along the Northeast Corridor, earned him traveling fellowships to visit France, Germany, and Japan in 1991 to study the land-use effects of the TGV, ICE, and Shinkansen system.

Mr. Gimpel is a member of the American Institute of Certified Planners, the American Planning Association, and the High Speed Ground Transportation Association. He currently works as a Senior Planner for Parsons Brinckerhoff in New York.

3 STATE-OF-THE-ART PRACTICES IN HIGH-SPEED RAIL RIDERSHIP FORECASTING: A REVIEW OF RECENT MODELING METHODOLOGY

Daniel L. Roth
Associate
Mercer Management Consulting
Washington, D.C.

3.1. INTRODUCTION

Over the past decade in North America, there has been a renewed interest in providing fast and reliable intercity passenger rail services. This interest has been driven by increasing traffic congestion on highways and at airports, in addition to environmental concerns. The success of new rail passenger systems in other countries has also fueled interest in North America. A variety of technological options have been proposed, from upgraded conventional rail services to new intercity "high-speed rail" (HSR) systems. Numerous proposals for specific service and technology scenarios have been considered in a variety of corridors and states.

In all cases, the starting point for engineering and financial feasibility analyses is an examination of the markets that are to be served. Revenue estimates; operating scenario and costs; rolling stock requirements and cost; stations; right-of-way, track and civil engineering structures and their cost – all of these aspects of a given project's scale and feasibility stem from a study of, first, the existing travel patterns in the corridor of interest and, second, of the ridership levels that can be expected for this new or upgraded intercity transportation service.

The objective of this paper is to present and critique the "state-of-the-practice" in the analytical methods used in forecasting ridership. After a brief section defining the nature and context of current HSR ridership modeling, the two-stage modeling system adopted in the most recent major studies of proposed HSR services is reviewed.[1] Overall strengths and weaknesses are summarized in the areas of data acquisition, disaggregate choice modeling, and forecasting methods. Finally, some areas of recent theoretical advances are suggested as holding some potential for improving the quality and reliability of HSR ridership studies.

This paper is concerned with modeling methodology; it does not review actual ridership forecast results across studies. A basic understanding of econometrics and model specification, estimation and testing is assumed. Some simplified definitions are given in sidebars to the text, but for a complete presentation of the underlying material, the reader is referred to such texts as Pindyck and Rubinfeld, 1991, and Ben-Akiva and Lerman, 1985.

3.2. MODELING ENVIRONMENT

This section reviews some important aspects of HSR ridership studies that shape the current modeling practice.

3.2.1. Intercity Mode Choice Modeling History

A few brief comments on the history of intercity mode choice modeling are worthwhile. (See Miller and Fan, 1991, and Ben-Akiva and Whitmarsh, 1984, for a more detailed review.) The first Northeast Corridor project of the late 1960s marked the beginning of serious attempts at these kinds of modeling systems. Early models were at a very aggregate level (e.g. gravity model), and they were first applied ignoring level-of-service (e.g., travel time). Eventually, the importance of including the level-of-service of competing modes was recognized, leading to sequential models which forecast the total travel in a given market given the mode shares. Aggregate models, however, have some fundamental weaknesses, including: the level-of-service data tends to be highly collinear, causing non-intuitive results and imprecision; the creation of large aggregation biases; the lack of a foundation in human behavior, in that it is the individual traveler who is the source of the O-D trips.

Starting in the late-1970s, the focus of forecasting for specific corridors and projects shifted to disaggregate models, which have their basis in the theory of individual consumers' utility functions. These models make efficient use of detailed travel data and provide estimation results superior to those of aggregate models. (See Ben-Akiva and Lerman, 1985, for the full theory of discrete choice models in transportation, as well as the detailed guidelines on their use.) The multinomial logit model, and subsequently the sequentially-estimated nested logit model (described in section 3.2), have found the most widespread use, and the numerous HSR ridership studies in recent years have relied almost exclusively on these models. An important implicit assumption in the use of these models is that in the future, a given type of traveler will have the same utility function that this type of traveler has now.

Common to the recent studies has been an integrated two-stage approach in the model system. A disaggregate mode choice model is estimated and used to forecast the aggregate market share for each mode between each O-D pair. Total travel between O-D pairs is forecast using an econometric travel demand model, to which the HSR share is applied to forecast HSR ridership. All the studies also follow the empirically-recognized necessity of segmenting the intercity travel market by – at a minimum – the business and non-business trip purposes.

Beyond these shared characteristics, however, each study has adopted a different overall modeling system, based on existing data availability, data-collection considerations, time and budget constraints, and the project context. In the studies, both revealed preference (RP) and specially-designed stated preference (SP) surveys are conducted. RP surveys gather and "reveal" data about current travelers, and are used to assess socio-economic characteristics and existing travel patterns. SP surveys solicit travelers' "stated" intentions about the use of proposed services such as HSR that do not yet exist in the markets being studied. Aggregate forecasts of ridership for each market segment are made separately using the two-stage approach. This modeling system is described in section 3.

3.2.2. Type of Rail Service

One must carefully distinguish in a specific geographic market between the two fundamental types of HSR service that are currently being proposed: improvement of existing rail services, and introduction of new rail services. In order to do this, the attributes of the service must be considered. If the attribute mix is sufficiently different from the attributes of the existing rail service, then the proposed service should be considered as an entirely new type of intercity service. Common sense serves well in this distinction.

The type of service, as defined above, has ramifications for the modeling system. Three scenarios can be identified:

a. *Improve existing rail.* As long as the attributes of the improved service are not too different from those of the existing service, one can rely on revealed preference data exclusively. The fact that the data is limited to the attribute range of the existing service is not likely to cause a significant bias in the forecasts. This is the least problematic scenario for a ridership study; no such HSR studies were examined, and this scenario is not considered in this paper.
b. *New service to replace existing.* In this scenario, because the new service does not yet exist, no revealed preferences in the range of the attribute mix are available. Stated preference data must be developed from surveys, and if possible combined with revealed preference data to take advantage of each data type's strengths.
c. *New service in addition to existing, or where none currently exists.* In addition to the concerns of scenario b. above, the appropriate nesting of alternatives in a nested logit model is less obvious. More extensive testing of the estimated model will be required, including both testing of the specification of model parameters and of the model structure (e.g. "independence of irrelevant alternatives," or IIA, tests).

3.2.3. *Institutional Environment*

It is important to recognize that the HSR ridership studies are performed in different institutional environments and for different types of clients. The studies have usually been performed by transportation engineering or econometric consulting firms. They sometimes consider one proposed HSR system, or a set of alternative projects with different attributes. The studies might be undertaken for a private group developing a HSR project, in which case they are critical to the financial feasibility and public perception of the project. Or they might be commissioned by a quasi-public authority, or a government agency at the state or federal level, in which case transportation and economic policy are key issues.

The ultimate use of the study has a non-negligible effect on: time and budget constraints on data collection and model estimation; the types of forecasts made as well as the level of aggregation; and finally the extent to which there exist pressures for the modeling results to be "adjusted" through various means to get more reasonable or attractive results. These institutional issues are beyond the scope of this paper, but they should be kept in mind whenever forecasting for HSR (or any transportation) services is undertaken.

3.3. MODELING SYSTEM

A generalized depiction of the two-stage modeling system is provided in Fig. 3.1. Note that the stage designation does not necessarily imply a required sequence. In fact, the first step is to identify available data, and gather new data using RP surveys and traffic counts, as well as SP surveys. Data on expected future levels-of-service on the different modes and future demographic and socio-economic characteristics must be collected as well. Then, as shown in the Exhibit, model estimation and forecasting are undertaken both for the total demand model and the mode choice model, and induced demand may be incorporated (discussed later).

The rest of section 3.3 reviews the key elements of the modeling system. The discussion is based on the findings of the review of recent HSR studies. Data acquisition issues are reviewed first, followed by comments on seven key aspects of the disaggregate choice models used in the studies. The section concludes with a shorter discussion of the total demand models.

3.3.1. Data Availability and Acquisition

This section briefly reviews three aspects of data availability and acquisition common to all the HSR ridership studies.

3.3.1.1. Limited Existing Data

All the HSR study efforts face a limited set of existing data. The existing data typically comes from sources such as: regional or state demographic studies, state highway counts, and the FAA 10% ticket sample. The existing data is used as follows:

- estimating the total demand models (in conjunction with the new RP survey data);
- determining the levels-of-service for the different modes at present and in the forecast years;
- as explanatory variables in the existing models to produce forecasts.

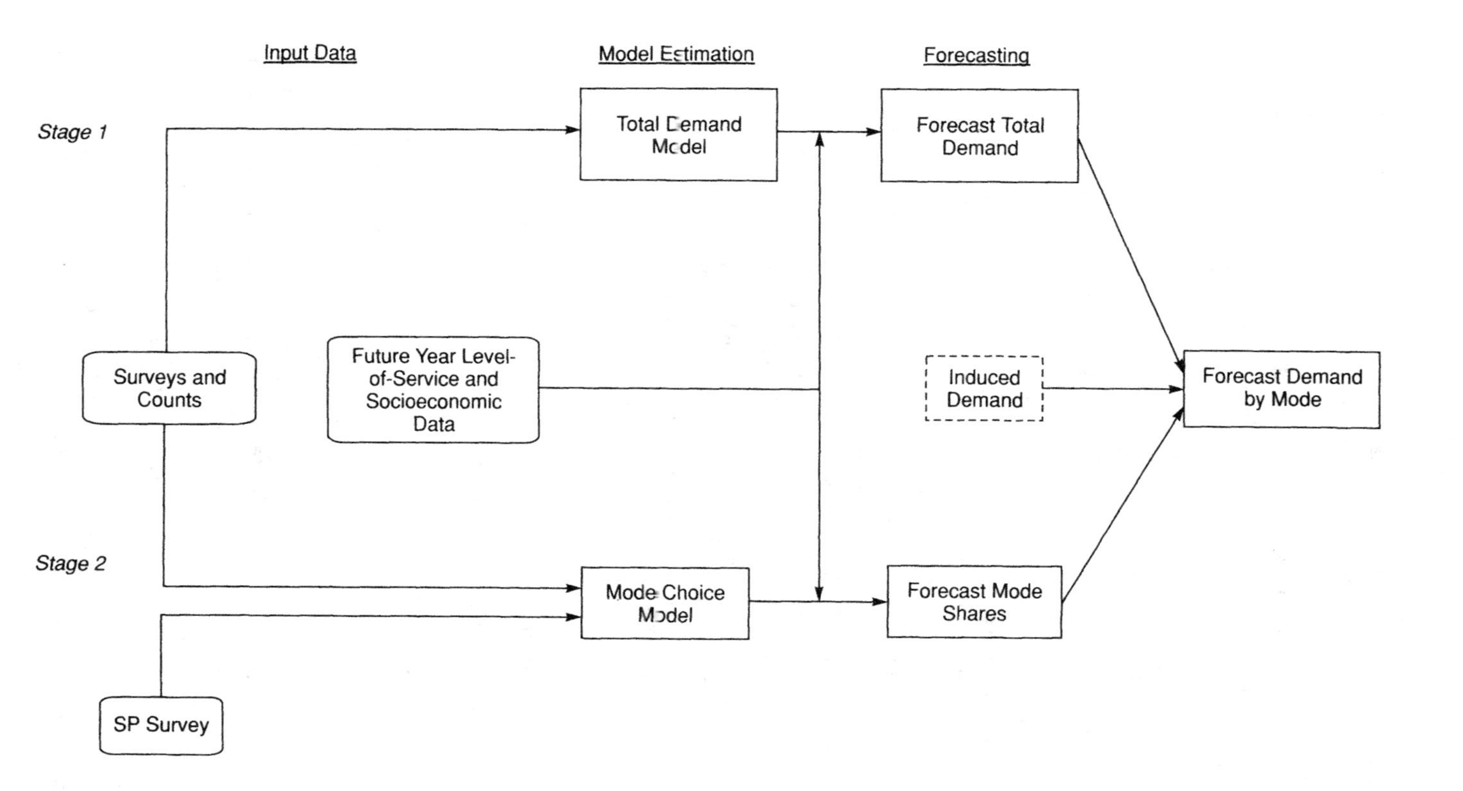

Figure 3.1. Two-Stage Modeling System.

The geographical origin/destination zones used in the studies are generally based on the zones predominant in the existing data (or fractions or multiples thereof, depending on data set size). In addition, because of a lack of easily accessible data sources, certain market segments with some likelihood of being tapped by HSR, such as intercity bus or tourist travel, have not been included in most studies.

3.3.1.2. Intercept Surveys

Both the RP and SP surveys are conducted by "intercepting" respondents as they travel. This is known more generally as choice-based sampling; that is, separate data samples are taken for each mode currently used, the travelers' current mode choice being thereby revealed. Choice-based sampling results in a loss of statistical efficiency, i.e. increased standard errors of the models' estimated coefficients. This is a result of the need to weight the sample data in order to retain consistency in the estimators.[2]

Balancing this efficiency loss is the fact that administering a survey for each mode separately is less costly than sampling the general population, say on a zonal basis, and ensuring sufficient representation of each mode. For the same cost, then, larger samples can be acquired using intercept surveys, which works to recover some statistical efficiency. Overall, intercept surveys allow a higher degree of control in terms of the questionnaire design and the survey administration.

3.3.1.3. Stated Preference Surveys

The use of stated preference techniques has become common. They are particularly useful in making predictions for services (like HSR) which do not yet exist, by estimating travelers' values of new service attributes. However, the way in which the new HSR services are presented must be carefully considered. Even with a simple and clear presentation, there is likely to be a certain amount of "justification bias" – the tendency for respondents to stick with their present mode choice regardless of the attributes of competing alternatives. On the other hand, if the presentation of HSR is rather flashy and its features overly stressed, an opposite "policy bias" may be created. The alternatives should all be presented in the same terms and with equal emphasis.

The SP surveys are typically based on a fractional factorial design.[3] The respondent is presented with a number of travel scenarios for which travel time, cost and other attributes are defined. In some surveys, the respondent then ranks the scenarios in order of attractiveness. In other cases, only two scenarios are presented at a time, and the respondent makes successive pairwise selections. These are the two most common approaches to assessing respondents' utilities (or disutilities) of various attributes, including travel time, service frequency, etc.

3.3.2. Disaggregate Choice Models

This section discusses key aspects of the disaggregate choice models as they were applied and estimated in the HSR ridership studies reviewed.

3.3.2.1. Logit Probability Function

The non-linear discrete choice probability function that has found unanimous use among HSR consultants is the logit function. The logit model is more popular than others (e.g., probit) because of its relatively simple theoretical foundation, its relatively low computational burden, and (perhaps most importantly) its ease of presentation and interpretation. The studies reviewed for this paper are no exception to this trend.

$$P_{in} = \frac{e^{V_{in}}}{\sum_{j \in Cn} e^{V_{jn}}} \qquad V_{in} = \beta X_{in}$$

The logit model has the general form shown above, where:

P_{in} = Probability of respondent n choosing mode i from a set of feasible alternatives Cn
V_{in} = Systematic utility of mode i for respondent n
X_{in} = Vector of explanatory variables (including attributes of mode i, and characteristics of respondent n)
β = Vector of model coefficients that are estimated

The logit function returns a probability of between 0 and 1, depending on the ratio of the mode i utility (numerator) to the expected maximum utility of all mode alternatives (denominator, also termed the "logsum" of the utilities). Note that the systematic utility function is assumed to be linear-in-the-parameters. The explanatory variables, however, can be non-linear combinations of other variables (e.g., ratio of travel cost to income) or "dummy" variables (e.g., 1 if owns car, 0 otherwise). Alternative-specific variables appear in the X vector only for a specific alternative.

Underlying the development of the logit model is the economic theory of consumer utility maximization. It is important to remember that utility is an ordinal, not a cardinal concept. Thus, using a ratio of utilities, which some studies claim makes intuitive sense, violates the underlying theory of the model, rendering meaningless any estimates obtained. More generally, it is also clear from the studies reviewed that even the relatively simple logit model is not always properly applied. Specific areas in which problems have been identified or that deserve more careful consideration are discussed in the sections below.

3.3.2.2. Choice Hierarchies

It is by now well-understood that a weakness of the earliest logit specification – the multinomial logit, with a single level of alternatives – is that the introduction of a new alternative mode would draw riders in equal proportions from all the other modes; this is the constant cross-elasticity characteristic of the multinomial logit model (or so-called "red bus-blue bus", or "IIA" problem). Some HSR studies have approached this problem by testing a variety of nested logit structures (in which alternatives in a given nest share attributes) and selecting one structure that provided the "best" fit to the data, presumably using standard IIA tests. This can be viewed as selecting the best-fitting cross-elasticity structure, i.e. estimating the degree of competition between modes.

Fig. 3.2 shows an example of a nested logit structure. The nested logit model is a generalization of the multinomial logit model: the multinomial logit model is estimated for each nest, and the choice probabilities passed "up" to the next-higher nest. Thus the choice probabilities of the alternatives in a nest are conditional on the choice probability of the entire nest, i.e., of the alternative from which it branches.

Other studies address the problem with different structures. In one modeling system, the HSR alternative is at the bottom of a nested logit structure consisting of only binary nests, so that the cross-elasticity of HSR does de facto vary with each other mode. However, this binary tree structure is clearly too restrictive: it does not allow for more than two alternatives in a nest. Another modeling system has as a primary motivation the avoidance of the constant cross-elasticity

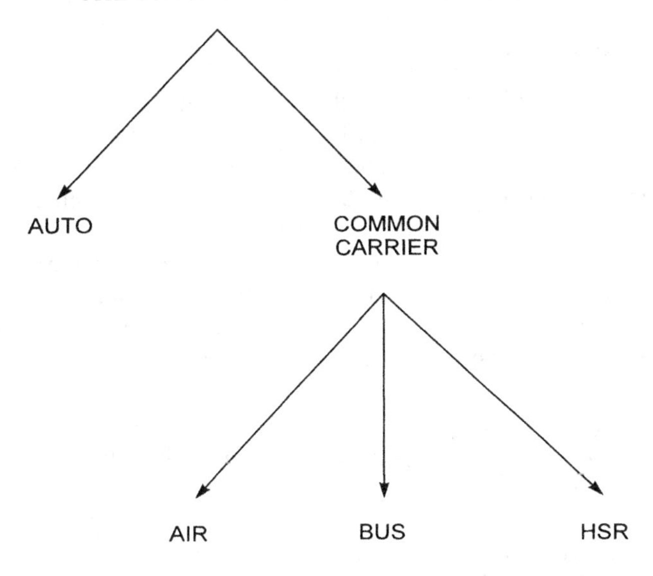

Figure 3.2. Example of Nested Logit Structure.

problem. It models diversion from each mode using a separately estimated model, thereby resulting in a unique HSR cross-elasticity with each mode.

It should be remembered, however, that the IIA characteristic of the multinomial logit model holds for the choice probabilities of each individual or group of individuals whose characteristics – as represented by socio-economic variables in the utility function – are homogeneous. In other words, for a heterogeneous traveling population, if the utility function is appropriately specified, there are only likely to be constant cross-elasticities within (and not between) each of that population's segments.

Overall, little uniformity is observed in the choice hierarchies modeled in the different studies, representing different approaches to the constant cross-elasticity property of the simple logit probability function. There is also limited evidence of thorough IIA testing, which can provide much insight into the statistical validity of the choice hierarchies modeled. Finally, only one study makes explicit mention that full-information estimation was performed.[4] Given the increasing availability of software capable of such estimation, nested choice hierarchies should no longer be estimated sequentially, a method which produces incorrect standard errors.

3.3.2.3. Market Segmentation

Examining the different mode choice and total travel behaviors of individual market segments can be accomplished either by estimating models separately for the different segments, or by including segment-specific variables in a single model. The basic market segments always found in intercity travel studies are the business versus non-business travel. Further distinctions are sometimes made for commute trips, multi-day trips, recreational trips, etc. In all the studies reviewed, there was at the very least a segmentation between business and non-business trips, accomplished by separate estimation of the models for each segment.

The market segmentations must not be arbitrary, but should be based equally on statistical tests and on reasonable hypotheses. Tests are available both to determine the significance of segment-specific variable coefficients, and to determine whether separately estimated models provide better results than a single estimation with pooled segments. Reasonable statistical testing of market segmentation should ideally be performed, despite the widespread a priori belief in the business/non-business segmentation.

It should be stressed that the reason we are concerned by the IIA property is that it may apply to each homogeneous market segment individually. It is not a concern for the homogeneous population as a whole (Ben-Akiva and Lerman, 1985).

3.3.2.4. Combining RP and SP Data

Some studies estimate the mode choice models using both the RP and SP data. The idea behind using both data types is that each has complementary strength and weaknesses. SP data is susceptible to a complex variety of biases. On the other hand, the SP survey is designed to ensure a greater orthogonality among variables, that is, so that the separate effects of normally correlated variables can be captured (e.g., high correlation in RP data between travel time and cost). A combined estimation, therefore, allows for SP data bias correction, and better precision in the estimation of variables that are typically correlated (Morikawa *et. al.*, 1991).

Each survey respondent is asked about his/her current trip (RP), and is then asked a number of questions based on hypothetical scenarios including HSR (SP). As a result, there are usually multiple SP data points for each RP response. For a number of reasons, it is not appropriate to simply weight each RP response by the SP multiple. Nor is it appropriate to estimate the model on a combined data set by simply pooling the RP and SP data, as some studies have done.

- First, weighting the RP data results in a loss of efficiency in the estimation (though consistency is retained, since this weighting is not choice-based). The standard errors from standard estimation software are therefore incorrect.
- Second, neither the correlation between the SP and RP responses, nor that between each respondent's SP responses, has been taken into account. To address the SP correlation, a non-parametric method could be used to estimate consistent standard errors: one common method is the Jackknife Method (first introduced by Tukey in 1958).
- More fundamentally, however, one may question the validity of the implicit assumption when simply pooling the data that the respective models underlying the SP and RP responses are the same. The attributes of the available alternatives, as well as a respondent's socio-economic characteristics, contribute to determine his or her utility for a given choice. The utility is latent; it is reflected in the respondent's revealed and stated preferences, but in different ways. When combining SP and RP data, then, two separate models must be considered. The RP and SP models have the alternatives' attribute variables in common, but the RP model also has variables that appear only in the RP data, reflecting characteristics that may affect actual choices made. Similarly, the SP model contains variables that may be included in the SP data, capturing potential biases (e.g. justification bias). A joint estimation of the RP and SP models should be undertaken; this will not involve weighting one of the data sets (Morikawa *et. al.*, 1991).

3.3.2.5. Specification of Variables in the Utility Function

For the mode choice models, the studies always test a great variety of utility function specifications, each with different combinations of explanatory variables. See Table. 3.1 for a typical final specification. The following explanatory variables are particularly often found in the studies (numbers correspond to the variables in the exhibit):

- Modified access/egress time (6). This variable is quite intuitively reasonable. In order to penalize access and egress times that are high relative to the trip length, the access/egress time is divided by total trip distance.

- Service frequency (8), which is incorporated using a transformation that increases utility as frequency increases, but at an exponentially decreasing rate. This is also a common and intuitively reasonable transformation: the marginal benefit of added service frequency decreases as frequency increases. Unfortunately, there is often no reasonable explanation provided in the studies for the different damping parameters selected. (Note that frequency is infinite for the auto mode.)
- Modified travel cost (5). Instead of travel cost alone, travel cost divided by income is tested as an explanatory variable. This produces a VOT which increases with income level, a reasonable situation. In the studies reviewed, however, modified travel cost never significantly improved the estimation results.

TABLE 3.1. Illustrative List of Common Explanatory Variables (Note that auto is base in this example.)

1	Air Constant
2	Bus Constant
3	HSR Constant
4	Total Travel Time (min)
5	Total Travel Cost/Income ($/$1000)
6	Access-Egress Time/Distance (min/mi)
7	Commuter: Access-Egress Time/Distance (min/mi)
8	Damped Frequency (1-exp(-k*freq)) (where k is the damping parameter)
9	Air: Distance
10	Group Travel Dummy

Another type of explanatory variable sometimes found and worth mentioning is the market-specific additive term. For example, within a model estimated only for business trips, a study might differentiate between commuter and non-commuter trips by including the modified access/egress time variable twice (6 and 7): once for all observations, and again only for observations for which the respondent was actually commuting between cities to get to work. The coefficient estimate for the first of the two variables (e.g., −2.1) is valid for non-commuters; the coefficient for commuters is determined by adding the coefficient estimates of the two variables (e.g., −2.1 + 1.3 = −0.8). The smaller disutility of modified access/egress time could reflect the fact that this portion of the trip is less onerous for commuters that for other business travelers who, say, may be on higher-valued company time. However, note that the difference in travel distance between these two groups is *not* a factor in the different coefficients, since distance is incorporated into the modified variable in the exhibit:

A final word of caution about the use of distance as an explanatory variable. As travel distance increases, air is usually favored over auto. But non-direct flights, which may for example involve deviation via a hub airport, result in longer air mileage without increasing the attractiveness of air travel relative to other modes. Thus, in order to avoid counter-intuitive results, the distance variable if used should be specific to air travel. Travel time, and not travel distance, is the appropriate generic (i.e., specified for all modes) explanatory variable to use in travel studies.

The inclusion of an air distance variable can thus be justified. The statistical significance and the sign of this variable should reflect the real-world non-linearity of travel time disutility as travel time increases. Specifically, in the range of travel times associated with air travel, say 90 minutes, additional increments of travel time (e.g., 5 minutes) are less onerous that those same increments are for shorter auto trips of, say, 30 minutes.

3.3.2.6 Value of Time

Travelers' value of time is fundamental to the analysis of the demand for transportation services. Indeed, *it is perhaps the most important factor* in an individual's choice of modes for a given trip, representing as it does the trade-off the individual is willing or able to make between travel time and travel cost. Consequently, the value of time must be considered carefully when estimating a mode choice model.

The value of time implied by a mode choice model estimation for a given market segment is calculated as the ratio of the travel time coefficient to the travel cost coefficient. Some HSR studies fix this ratio a priori for the different market segments. This is not in and of itself incorrect *if the values thus defined can be justified*, for example on the basis of other recent studies in corridors with similar demographic and employment characteristics. Yet it involves the strong assumption that the population in a given market segment is homogeneous with regard to its value of time. The studies that do pre-select their values of time tend to inadequately document how the values were chosen.

A better approach, taken in other studies, is to determine the "average" value of time for a market segment *based on the estimated coefficients*. This value should then be discussed and compared to values determined for other segments and in other studies. If separate modal diversion models are estimated (see next section), a separate value of time can also be associated to users of different modes in the same market segment (e.g. auto business VOT vs. air business VOT).

The travel time and travel cost variables can be interacted (i.e. multiplied or divided) with socio-economic indicators if desired, in order to asses value of time variation across sub-segments of the population. In addition, separate values of time for access/egress, waiting and line haul time can be identified if travel time is so divided in the model. Finally, some studies also calculate the value of service frequency.

Table 3.2 indicates the wide range of values of time pre-selected or determined in the studies. Given the limited number of studies presented, and the variability in the quality of the methodologies, the table is not meant to illustrate a definitive reasonable range of VOT's for future studies. Furthermore, each study is based on a different population with different characteristics. What is clear instead is that it is very difficult to compare values of time across studies, as each study uses a different utility function specification.

TABLE 3.2. Range of VOT Results in Recent HSR Studies ($/hour).

Study	*Business*	*Non-business*	*Remark*
a.	15 -45	6 -14	Varying by income, from $20,000-$60,000.
	86 -20	18 -4	Value of access/egress time, varying by total distance, from 90-380 miles.
b.	35, 20	28, 9	For air and auto, respectively.
	24, 13	19, 6	Access/egress time, fixed at 2/3 line-haul time, for air and auto, respectively.
c.	65, 43, 40	34, 26, 28	For air, auto, and rail, respectively.
d.	60, 40	30, 15	Pre-selected VOT values. Business: single-day & multi-day. Non-business: non-recr. & recreational.
FAA	1.0 to 1.5 × average wage		Based on comprehensive review of studies.[5]

3.3.2.7. *Diversion Models*

Diversion models are a series of binary discrete choice models where the alternatives are, in each case, an existing mode and HSR. The models are estimated separately using the separate choice-based SP data. In this way each mode is really being treated as an individual market segment from which HSR should attract riders. The two-stage modeling system of Figure 3.1 is essentially repeated, each time using the data sample for one existing mode. Two motivations are typically given for this approach:

- HSR services are not likely to divert travelers equally from existing modes. Thus, as discussed earlier, if population heterogeneity is not accounted for in the utility function specification, the resulting constant cross-elasticities of the multinomial logit model may be inappropriate. Diversion models allow each mode to have a unique cross-elasticity with HSR. (Note, however, that a properly structured and tested nested logit model also allows for different cross-elasticities.)
- Travelers' mode choices are a reflection of their value of time, as well as the attribute mix of the alternative services. For example, in intercity markets air will tend to serve travelers with a higher value of time than auto, all else being equal. Auto travel, as a private mode, provides unique features such as flexibility of departure and itinerary, availability for use at destination, etc. Consequently, the modeling system should allow for the development of separate functional forms and variable specifications for each existing mode's market segment.

Despite these features, there are some limitations to be aware of:

- Because each of the separate models estimated only accommodates diversion to HSR, the use of diversion models makes a strong implicit assumption that there will be no mode switching behavior between existing modes once HSR is introduced. Put differently, the assumption is made that, for instance, the air and auto markets are independent. This assumption is reasonable only if the price and level-of-service of the existing modes remains unchanged. Yet this may not hold in cases where HSR systems add a relatively large transportation capacity to the intercity travel market.
- There is by design no opportunity with diversion models to calibrate the mode share forecasts to current mode shares. (See the section on Ridership Forecasts below.)

Finally, some studies are able in another fashion to incorporate different diversion (or "switching") behaviors for travelers using different existing modes. This is accomplished by specifying the usual set of alternative-specific constants repeatedly for each data segment defined by a current mode. Thus, there might be three (three existing "source" modes) times three (three alternative modes, minus the base alternative, plus HSR) equals nine constants. This method, however, is not as rich as the use of diversion models in that instead of more tangible characteristics such as value-of-time, only ill-defined "inherent" characteristics captured in the constants can differentiate the current travelers.

3.3.3. *Total Demand Models*

This section briefly presents the application of total demand models in the modeling systems.

3.3.3.1. *Total Demand Models*

The studies all tend, after appropriate testing, to select a model specification in which the dependent variable, total interzonal or intercity trips, is a multiplicative function of a number of explanatory variables. Taking the logarithm (typically to the base *e*) of the multiplicative model results in a linear form, shown below, that can easily be estimated. This log-transformed specification happens to be easy to understand, since each coefficient can be interpreted as the elasticity of demand to its respective explanatory variable.

$$\ln (T_{ij}) = \alpha + \beta_1 \ln(X_1) + \beta_2 \ln (X_2) + \ldots \beta_k \ln (X_k)$$

where T_{ij} = Number of trips between zones i and j
X_k = Explanatory variables (socio-economic, level-of-service, etc.)
α, β_k = Parameters

Because of the multiplicative relationship between the variables, however, there is the risk when forecasting demand of exacerbating the effect of relatively small errors in the forecasted independent variables that are used along with the estimated parameters. Indeed, the Northeast Corridor experience with this model formulation has not been very positive (Ben-Akiva and Whitmarsh, 1984). Nevertheless, there is a certain momentum behind the use of this specification: it is a simple relationship, and it has been used consistently over time, thus giving it the legitimacy of experience with consultants. The elasticities that it identifies can (and should) be compared to elasticities estimated in other studies. That being said, the final specification of variables for this model still varies greatly between studies, with a variety of socio-economic characteristics and mode attributes being used.

The log-log model described above is sometimes applied during forecasting using variables which are the ratio of the future to base year values, i.e. variables representing growth. Such a growth formulation may serve to lessen any forecast errors, since it provides an incremental forecast over the actual observed travel demand. Finally, if diversion models are being used, separate log-log regression models must be used for total trips for *each* mode. If travel cost or fare is included as an explanatory variable, this approach has the advantage of allowing an examination of the different price elasticities of travel for each mode.

3.3.3.2. Induced Demand

Induced demand is the set of additional trips that are made on the new mode because of its higher level-of-service (or improved "value for the money"). This demand is in addition to new demand due to normal growth. Studies may explicitly or implicitly include induced demand in the modeling system (hence the dashed-line representation in Figure 3.1). There is thus far little uniformity in the specific approach used.

Explicit induced demand estimation assumes that the expected maximum utility (logit's logsum denominator) after HSR introduction will be greater than that before, so that travel frequency will increase. The determination of the level of induced demand may involve the use of a model of the form shown below. If β is not given an a priori value, such as 1, then the model must be estimated, after having collected the appropriate SP data on travelers' willingness to make additional trips after HSR introduction.

$$\ln\left(\frac{Freq_1}{Freq_o}\right) = \beta \ln\left(\frac{Sum_1}{Sum_o}\right)$$

where $Freq_0$, $Freq_1$ = Trip frequency before and after new service introduction

Sum_0, Sum_1 = Expected max. utility (logit's logsum denominator,
see section 3.3.2.1) before and after

β = Parameter

Once the parameter β is estimated or selected, the ratio of travel frequency can be calculated for a given HSR scenario, and then used as a multiplier of forecasted HSR trips. This induced demand model is conceptually sound, but note that it relates changed *HSR* ridership to *HSR* utility only (in that the expected maximum utility differs only in the addition of HSR travel utility) and not the utilities of other modes. In addition, though the model could be estimated separately for the trip purpose segments, it fails to include any other socio-economic variables to account for people's different responses to the increase in utility.

Alternatively, the trips induced by HSR can be incorporated implicitly if the total demand model includes a variable that represents the level-of-service attributes (particularly travel time and cost) of the new mode. For example, the logsums may be included as explanatory variables in the total demand models.

3.3.3.3. Ridership Forecast

Before HSR ridership forecasts are finally calculated, those mode choice models involving more than one existing mode (i.e. not the diversion models) are calibrated: modal constants are adjusted so that use of the aggregate base year data results in the actual base year shares. Studies normally rely heavily on such calibration, since the forecasts will have a better correspondence with reality. In the case of diversion models, no such calibration is possible at all. Of course, *all* models have an Achilles heel in that no calibration is possible for the HSR share, since there is no current share to examine. Among the studies reviewed, there appear to be cases where additional, perhaps inappropriate, and in any case vaguely justified changes are made to the modal constants.

In general, as was indicated in Figure 3.1, the forecast of HSR ridership is calculated by applying the forecasted HSR mode share, based on the estimated mode choice model, to the forecast future total demand, based on the estimated total demand model. Induced demand multipliers are applied where they exist. This procedure is repeated for each market segment for which models were separately estimated. In the case of diversion models, the fraction of HSR diversion from a given mode is applied to the future demand forecast for that mode, and total HSR ridership calculated as the sum of the trips diverted, plus those induced.

3.4 NEW DIRECTIONS

This review of HSR studies has focused on issues of basic and contemporary theory, sampling and model design. There are well-established areas of more advanced econometric and choice modeling theory which could be considered for future application to HSR ridership modeling. Three of these advanced areas are briefly suggested here. (A full presentation is beyond the scope of this paper; references are provided.)

3.4.1. Latent Variable Models

HSR offers a travel environment, amenities and "experience" that is different in many ways from that of any other mode. It would be useful to try to capture latent (i.e., unobservable) factors of intercity mode choice, such as perceptions of comfort, travel environment and amenities, and to incorporate them in the modeling process. The importance of these factors in a traveler's choice of modes is intuitively clear, especially given experience with actual HSR systems in other countries. Supporting this intuition is the fact that the estimated choice models almost always include a significant modal constant which, absent any other measures, is interpreted as reflecting travelers' "inherent" (i.e., latent) preference for the given mode. Furthermore, we rely on SP surveys which frequently include a description of the new rail service beyond travel time, frequency and travel cost alone. This description, which gives the respondents a feel for the new mode, certainly has an effect on their stated preferences.

Latent variable models would be used to estimate a number of latent variables of travel, which would subsequently be incorporated into the utility function of the choice model. For example, one could hypothesize three latent variables as follows: comfort (related to seat design, leg room, etc.); amenity level (related to provision of food service, phone and fax services, audio-visual equipment, etc.); and environment (related to seat and table arrangement, lighting, background noise, etc.). A survey would need to be designed that would present the features mentioned above (e.g., leg room, food service, noise level), and through some sort of rating, comparison or other scaling scheme, would develop a data set of indicators of the latent variables. Given the features and the indicators, then, the latent variables themselves could be estimated.

An argument could be made that instead of going through the process of modeling the latent variables, one could directly incorporate generic modal features as discrete or dummy explanatory variables in the utility function. The motivation for using latent variables, however, is that the success of a mode's features in attracting riders is not a *deterministic* function of the features' provision. Rather, what must be modeled are the *perceptions* of the quality of the overall level-of-service, to which the features contribute.

See Bollen, 1989, on latent variable models. For recent applications, see for example Ben-Akiva and Ramaswamy, 1993, and Gopinath and Ramaswamy, 1993.

3.4.2. Transfer of Results

The transferability of results refers to the ability to use an estimated model from a previous study to enhance a new one. If, for a new study or a new base case scenario, a sample is not available, it may be desirable to simply update ("calibrate") a preexisting model. In this case, calibration is performed using ad hoc methods, for example, plugging in aggregate explanatory variables and adjusting the modal constants to get the base year shares. The trade-off here, however, is high forecasting errors due to the aggregation involved. If a sample is available, a similar approach can be used which involves adjustment of the model's constants until the aggregate predicted shares match the observed shares. Neither of these approaches are satisfactory.

Alternatively, in some cases a model's parameters can be properly estimated, but time or budget constraints are such that only a relatively small sample was collected. The parameters from a prior model can be incorporated to enhance the efficiency of the new model. Techniques to this end have been developed in the literature; see for example Ben-Akiva, 1980, Gunn *et. al.,* 1985, and Ben-Akiva and Bolduc, 1987. Methods include the use of a "transfer scaling estimator" and a "combined transfer estimator," which seek to include prior estimators in the current model to reduce variance, while controlling for bias due to this transfer of data. The requirements and complexity (i.e., cost) of the methods vary, and the selection of a method depends, among other things, on data availability and the quality of the so-called "donor" model.

The point here is that many of the HSR ridership studies are conducted in the context of conceptual project development or preliminary planning and engineering. As such, there is a limited amount of time and funding available. Occasionally, however, a study may be performed for which a much more extensive and rich data set is acquired. The results of such a study could be used to enhance the results of a number of other smaller-scale studies in a "trickle-down" effect. Ideally, an effort would be made to centrally collect HSR studies, their data and modeling results. One would then begin by simply comparing coefficients for similar variables across the larger "donor" studies. Values of time and elasticities would also be examined. One could further estimate different model forms using samples from different studies, and compare the results. Such comparisons would ultimately help produce more reliable ridership forecasts and enrich the "knowledge-base" for HSR travel demand and mode selection forecasting.

3.4.3. The Probit Model of Discrete Choice

An alternative to the logit model, in which the error terms are assumed to be distributed "iid Gumbel", is the probit model, in which we assume that the error terms take on a normal distribution. A major advantage of the probit model is that it allows for the use of any particular variance-covariance structure. What this means in practical terms is that independence between alternatives is not required and that specific correlations can be incorporated in order to represent shared unobserved attributes between the alternatives. Thus, nesting is not required, and the constant cross-elasticity problem is avoided. Furthermore, it is relatively easy to incorporate random taste variations and other relaxations of the model. In other words, overall, probit has a more general structure, with which one can test a greater variety of underlying behavioral hypotheses.

The probit model has been in use for over a decade, and is covered in many texts, such as Daganzo, 1980, and Hausman and Wise, 1978. The main reason probit has not been extensively used in demand modeling – and not used at all in HSR studies – is that it is very computationally burdensome, as its functional form is expressed in terms of an integral. In recent years, however, application of numerical methods to maximum likelihood estimation has been refined, as well as sped up with the increase in computer power. Thus its use may soon become more common. In considering whether to use the probit model, then, there is a trade-off between modeling flexibility and computational cost which must be carefully weighed. In fact, a well-specified full-information nested logit estimation may be quite adequate for modeling in preliminary studies of the kind reviewed for this paper.

3.5. SUMMARY AND CONCLUSION

The review of recent HSR studies reveals that the modeling systems in use share a common basic structure. There is uniform reliance on the disaggregate logit model as part of a two-stage approach combining mode choice and total demand models. However, the actual logit nested structure, utility function specifications, total demand models and incorporation of stated preference data all vary greatly, and problems are often found with the methodologies. For example: there is little evidence of statistical testing of the IIA assumption or of market segmentation; standard errors are incorrect, exacerbated by ad hoc weighting schemes; there is little support for pre-selected traveler values of time. Since the ridership forecasts for a given HSR project are critical to its planning and engineering, as well as its financial feasibility, erroneous forecasts naturally increase the chances of financial, operational or capacity problems in the future.

To be fair, the studies are done in the context of *preliminary* planning and engineering; later more detailed studies should benefit from more time and larger budgets. Nevertheless, the major problems identified within the studies are worrisome, for they may be repeated in the future. It is certainly true that one should view with great skepticism forecasts that do not appear reasonable based on current or past experience (e.g., failing the so-called "giggle test"), even if the forecasts are obtained from sophisticated models. On the other hand, a reasonable forecast should not validate a modeling system as appropriate for further use if that modeling system is inherently flawed. In light of this, it is worth noting that some modeling systems that on the surface might be considered the least "sophisticated" may in fact be favorable, since they are the least complex (i.e., error-prone) in terms of the use of SP data and the estimation of models, and they provide simple and clear results.

It is encouraging to see that stated preference survey techniques are finding widespread use; they are particularly important since HSR has unique attributes, and does not yet exist in the geographic areas covered by the studies. Yet while many consultants have turned to using the "latest" in stated preference techniques and sophisticated (as well as time-saving) PC-based surveys, sufficient resources and care must be allocated to properly combining the RP and SP data, and incorporating the data into the modeling system.

These points are all a manifestation of a larger problem, namely that the modeling systems in the studies are often treated as proprietary, resulting in a lack of peer review and informed debate in this niche industry. Ultimately, this situation will need to be addressed, if decisions regarding transportation policy and resource allocation are to be effective. One possibility would be the creation of a central "laboratory" where samples and results from studies across the U.S. could be compared and analyzed, perhaps in conjunction with the Federal HSR funding program recently proposed.

There are several ways, such as the three suggested in this paper, in which the frontiers of econometric theory and practice could be applied to the task of modeling HSR ridership. Of course, this will complicate the consultants' work, requiring more time and money, and making communication of the modeling process to clients more difficult. Yet computing power and speed continue to grow and become less expensive, and software packages are becoming more sophisticated, with capabilities for some of these advanced methods. For the time being, however, it would be wiser to focus on the fundamentals and iron out the existing problems that exist in some of the consultants' modeling systems. The consultants are clearly working under a variety of pressures and for different types of clients, but the basics must be properly understood and applied.

Of course, forecasting for investment in other modes is not always up to these high standards either. One could argue that it is not fair to ask so much of HSR studies, and not of other modes that also compete for public funds and public support. The answer is to critique the other studies and demand higher standards of them as well. After all, in many cases, the same consultants do those studies too. HSR is a contentious issue in North America, and because of this and its lack of history here, it will continue to be closely scrutinized. Overly optimistic projections that burden a specific project may hurt the mode's possibilities more generally. While recent HSR ridership studies are good pieces of work, their methodologies can be further improved upon in future studies. It is imperative that this alternative mode of transportation, which holds so much potential for addressing intercity travel needs in the North America, be given fair consideration using sound methods of analysis.

FOOTNOTES / ENDNOTES

1. The Final Reports of five specific North American HSR studies betwen 1990 and 1993, by bour different consultants, were used in this review.
2. Weights are, however, not required if the multinomial (and not nested) logit model is being used. In such case, all that is required is an adjustment of the alternative-specific constants. (Ben-Akiva and Lerman, 1985).
3. Each level-of-service and travel cost attribute can take on different discrete levels (eg. half-hourly service, hourly service and bi-hourly service). A full factorial survey design would present as choices all combinations of the different attribute levels. The fractional factorial design selects, based on statistical considerations, a subset of these combniations that provides the most usedful trade-off information.
4. In full-information estimation of nested logit model, all choice levels of the model are estimated simultaneously. Until recently, statistical software was only capable of multinomial (i.e. one-level) logit estimation, requiring a nested model to be estimated one nest at a time.
5. Federal Aviation Administration U.S. DOT, June 1989. Economic Value for Evaluation of Federal Aviation Administration Investment Regulatory Programs.

REFERENCES

Ben-Akiva, Moshe. "Issues in Transferring and Updating Travel Behavior Models." In Stopher, P.R., A.H. Meyburg and W. Brog (eds.), New Horizons in Travel Behavior Research. Lexington, MA: Lexington Books. 1980.

Ben-Akiva, Moshe and Denis Bolduc. "Approaches to Model Transferability and Updating: The Combined Transfer Estimator." Transportation Research Record, no. 1139, 1987, pp. 1-7.

Ben-Akiva, Moshe and Steve Lerman. Discrete Choice Analysis: Theory and Application to Travel Demand. Cambridge, MA: The Massachusetts Institute of Technology Press. 1985.

Ben-Akiva, Moshe and Takayuki Morikawa. "Estimation of Travel Demand Models from Multiple Data Sources." Transportation and Traffic Theory, 1990, pp. 461-476.

Ben-Akiva, M., and Ramaswamy, R. "An Approach for Predicting Latent Infrastructure Performance Deterioration." Transportation Science, vol. 27, no. 2, 1993.

Ben-Akiva, M.E. and C.C. Whitmarsh. "Review of U.S. Intercity Passenger Demand Forecasting Studies." In Proceedings of the International Seminar on the Socio-Economic Aspects of High-Speed Trains, Paris, November 5-8, 1994.

Bollen, K. A. Structural Equations with Latent Variables. John Wiley & Sons. 1989.

Brownstone, David and Kenneth Small. "Efficient Estimation of Nested Logit Models." Journal of Business & Economic Statistics, Vol. 7, No. 1, 1987, pp. 67-74.

Daganzo, C. Multinomial Probit. New York: Academic Press. 1980.

Daly, Andrew. "Estimating 'Tree' Logit Models." Transportation Research, 21B, No. 4, 1987, pp. 251-67.

Federal Aviation Administration. Economic Value for Evaluation of Federal Aviation Administration Investment and Regulatory Programs. U.S. DOT, June 1989.

Forinash, Christopher. A Comparison of Model Structures for Intercity Travel Mode Choice. Master's thesis. Evanston, IL: Northwestern University. May 1992.

Gopinath, D., M. Ben-Akiva and R. Ramaswamy. "Modeling Performance of Highway Pavements." Presented at the 72nd Annual Meeting of TRB, Washington, DC, January 1993. Forthcoming in Transportation Research Record.

Gunn, H.F., M. Ben-Akiva, and M.A. Bradley. "Tests of the Scaling Approach to Ttransfering Disaggregate Travel Demand Models." Transportation Research Record, no. 1037, 1985, pp. 21-30.

Hausman, J., and D. Wise. "A Conditional Probit Model for Qualitative Choice: Discrete Decision Recognizing Interdependence and Heterogeneous Preferences." Econometrica, vol. 46, 1978, pp. 403-426.

Hensher, David. "Sequential and Full-Information Maximum Likelihood Estimation of a Nested Logit Model." The Review of Economics and Statistics, Vol. 68, No. 4, 1986, pp. 657-67.

Koppelman, Frank. "Predicting Transit Ridership Response to Transit Service Changes." Journal of Transportation Engineering, Vol. 109, No. 4, July 1983, pp. 548-64.

Miller, Eric and Kai-Sheng Fan. "Travel Demand Behaviour: Survey of Intercity Mode-Split Models in Canada and Elsewhere." Directions, The Final Report of the Royal Commission on National Passenger Transportation, Vol. 4. Ottawa, Canada: Ministry of Supply and Services. 1992.

Morikawa, Takayuki, Moshe Ben-Akiva and Kikuko Yamada. "Forecasting Intercity Rail Ridership Using Revealed Preference and Stated Preference Data." Transportation Research Record, No. 1328, 1991, pp. 30-35.

Pindyck, Robert and Daniel Rubinfeld. Econometric Models & Economic Forecasts. New York, NY: McGraw-Hill, Inc. 1991, 3rd edition.

Tukey, John. "Bias and Confidence in not quite Large Samples." Annals of Mathematical Statistics, 29, 1958.

NOTE:

Five different major high-speed rail studies conducted between 1990 and 1993 were reviewed. They were managed by four prominent consultants. The studies focused on a total of four corridors: three in the U.S. and one in Canada. They are not referred to specifically, as the intent of this paper is not to critique them individually, but to raise awareness about strengths and weaknesses in the key elements of the modeling methodologies used in HSR studies.

DANIEL L. ROTH

Daniel L. Roth is an Associate with Mercer Management Consulting in Washington D.C. He provides strategic management consulting and economic and financial analysis for transportation clients, covering infrastructure, operations, and logistics. His transportation expertise spans market forecasting, cost-benefit analysis, financial planning, and strategic system planning. He experience includes positions at Price Waterhouse LLP, the Federal Railroad Administration and local agencies, as well as in rail transportation in Europe. A particular focus has been the development of innovative public-private partnerships for transportation infrastructure and system. He has been involved with the development of high-speed transportation systems for the past 8 years, has authored several papers, and has been selected to be the Chairman of the Transportation Research Borard's Committee on Intercity Rail Passenger Systems. Mr. Roth holds a Masters Science in Transportation from the Massachusetts Institute of Technology and a bachelor of Science in Systems Engineering from the University of Pennsylvania.

4 NOISE/VIBRATION AND ELECTRIC AND MAGNETIC FIELDS OF HIGH SPEED RAIL/MAGLEV SYSTEMS

Carl E. Hanson, Ph.D., P.E.
Harris Miller Miller & Hanson Inc.
15 New England Executive Park
Burlington, MA 01803

4.1. NOISE AND VIBRATION

The introduction of a new transportation system generates concerns about the change in the noise and vibration environments brought about by the new source. When the new source has unique features, as in the case of high speed rail or maglev systems, or when the community has not had prior exposure to a particular source, the concerns are heightened. The classic example is the introduction of the first truly high speed rail system, the Shinkansen in Japan starting in 1964. Public reaction to noise and vibrations from the new trains resulted in new standards in that country and major efforts were undertaken to preserve the environment. Great strides have been made in understanding the nature of noise and vibration sources from guided surface transportation systems since the early days of the "Bullet Train." A great deal of research on sources of noise and vibration from high speed trains has resulted in systems that are well adapted to the environment.

This section presents the key acoustical terms commonly used in descriptions of noise and vibration environments, summarizes the noise and vibration characteristics of high speed rail/maglev systems, provides the criteria used to assess noise and vibration impacts in various countries, demonstrates an approach for assessing impact, and discusses countermeasures against noise and vibration from high speed rail/maglev systems.

SIDEBAR: Basic Accoustical Terminology

The sounds that we hear are the result of very small pressure fluctuations in the atmosphere around us. In order to describe the signal content of these pressure fluctuations, acousticians have developed methods of analysis that differentiate among loudness, pitch and time history of sound. This sub-section is intended as a brief introduction to the descriptors to be used in this chapter. More detail can be found in an acoustical text or noise control handbook.[1] Although some authors take care to define them separately, throughout this report we use the terms "sound" and "noise" interchangeably.

Noise Level, Decibels

Sound is a description of pressure oscillations above and below the mean atmospheric pressure. The amplitude of oscillation is related to the energy carried in a sound wave; the greater the amplitude, the greater the energy, and the louder the sound. The mean value of the pressure oscillations is always the atmospheric pressure; consequently, to describe an effective value of sound pressure, we use the root mean square pressure. The full range of sound pressures encountered in the world is so great that it becomes more convenient to compress the range by the use of the logarithmic scale, resulting in one of the fundamental descriptors in acoustics, the **sound pressure level, (Lp)**, defined as:

$$L_p = 20\log_{10}(p/P_{ref}),\ \text{in decibels (dB), where}$$

p is the sound pressure and P_{ref} is the reference sound pressure, internationally adopted to be 20 micropascals. In this chapter, the term **noise level** also refers to the sound pressure level, Lp.

Frequency Spectrum, A-Weighting

In the discussion above, we relate noise level to the amplitude of pressure oscillations. Another aspect of the oscillation is its frequency, the number of complete cycles above and below the mean value that occurs in a unit time. The unit is cycles per second, called Hertz (Hz). When a sound is analyzed, its energy content at individual frequencies is displayed over the range of frequencies of interest, usually the range of human audibility from 20 Hz to 20,000 Hz. This display is called a **frequency spectrum**.

Sound is measured using a sound level meter, with a microphone that is designed to respond accurately to all audible frequencies. On the other hand, the human hearing system does not respond equally to all frequencies. Low frequency sounds below about 400 Hz are progressively and severely attenuated, as are high frequencies above 10,000 Hz. To approximate the way the human interprets sounds, a filter circuit with the same frequency characteristics as the typical human hearing mechanism is built into sound level meters. Measurements with this filter enacted are referred to as **A-Weighted Sound Pressure Levels**, expressed in **dBA**. Sounds at frequencies below 20 Hz (infrasound) and above 20,000 Hz (ultrasound) are generally imperceptible by the human hearing system and are consequently neglected in an acoustical analysis.

Noise Descriptors: Lmax, Leq, SEL and Ldn

Another characteristic of sound in the environment is its fluctuation in level over time. Several descriptors have been developed to provide single number metrics for these variations. As a vehicle approaches, passes by, and recedes into the distance, the sound pressure levels rise and fall accordingly. Although detectable at levels slightly lower than the background sound level, the passby event is considered to occur over a duration containing most of the sound energy, such as within 10 dBA or 20 dBA of the peak.

The descriptor used for representing the highest sound level of a single event, such as the passby of a maglev vehicle , is the **Maximum Level, Lmax**. Lmax in dBA is commonly used to compare noise levels from different vehicle passbys, but it is important to understand that unless the sound is steady and continuous, the maximum level occurs for only a short time during an event. It is usually dominated by the single loudest source, which may be only one vehicle in a long train. A shortcoming of Lmax is that it ignores the duration of the event, an important environmental consideration. A single event descriptor that accounts for both level and duration of a sound is the **Sound Exposure Level, SEL**, which is a single number unit in decibels that describes all the sound energy received at a given point from an event, but normalized to a one-second duration. The normalization to one second allows comparison of the sound energy, and eventual combination, of different types of events on a common basis. For example, the SEL can be used to compare the sound energies emitted by various kinds of trains, even if they have different lengths.

The descriptor used for cumulative noise exposure in the environment is the **Equivalent Sound Level, Leq**. This is the level of a steady sound which, over a referenced duration and location, has the same A-weighted sound energy as the fluctuating sound. The duration of one hour is commonly used in environmental assessments. Researchers in Germany often describe train noise by the "passby level" which is the Leq over the time it takes for the train to pass in front of a given point. The "passby level" is typically somewhat lower than the actual Lmax because it is less influenced by a single dominant source. Environmental impact assessments in the United States use the **Day-Night Sound Level, Ldn**. Ldn is a 24-hour Leq, but with a 10 dB penalty assessed to noise events occurring at night during the hours of 10 pm to 7 am. Ldn has been found to correlate well with the results of attitudinal surveys of residential noise from transportation sources. It is the designated metric of choice of many Federal agencies, including Department of Housing and Urban Development (HUD), Federal Aviation Administration (FAA), Federal Transit Administration (FTA), Federal Railroad Administration (FRA) and Environmental Protection Agency (EPA).

Vibration Descriptors: Lv

Vibration is a rapidly fluctuating motion with an average of motion of zero. For example, a building floor may shake when a train passes, but in the end, the floor stays in one place. An exception, of course, is the extreme cases where vibrations are so severe that permanent damage occurs, but this does not occur from train passbys. Train vibrations are generally described by the effective velocity level, determined by the root-mean-square (rms) of the velocity of the motion and expressed in terms of decibels with a reference level of one microinch per second in the U.S. (Other countries use either 1×10^{-8} meter/second or 5×10^{-8} meter/second as a reference.) We use VdB to differentiate vibration decibels from noise decibels.

4.1.1. Noise and Vibration Characteristics

Noise and vibration characteristics of high speed ground transportation systems have been determined through extensive measurement programs conducted by researchers in many countries.[1,2,3] Noise measurements are made with microphones placed at a specific reference distance from the tracks (100 feet in the United States and 25 meters elsewhere) and a standard height over a datum plane (5 feet above ground surface in the United States and either 1.5 meters above ground or 3.5 meters above top of rail elsewhere). Usually the reference data are presented in terms of maximum noise level (Lmax), Sound Exposure Level (SEL), or Passby Level (Leq,p), depending on how the data will be used.

Special microphone configurations are used in diagnostic work, such as lines of microphones at various distances to determine sound propagation effects, or arrays of microphones used to locate sound sources on the moving trains. Data from these microphones are analyzed in terms of noise level and frequency spectrum to gain information about how each noise source changes with operating conditions, speed, track conditions, and power settings.

Noise level from high speed trains increases exponentially with speed in a fairly predictable manner. For electrically- powered trains, wheel/rail noise level generally increases as the third power of velocity up to a transition speed where aerodynamic noise sources on the body surface begin to dominate; above the transition speed the noise level increases as the sixth power of speed. Figure 4.1 shows the general relationship of noise increase with speed, indicating the transition

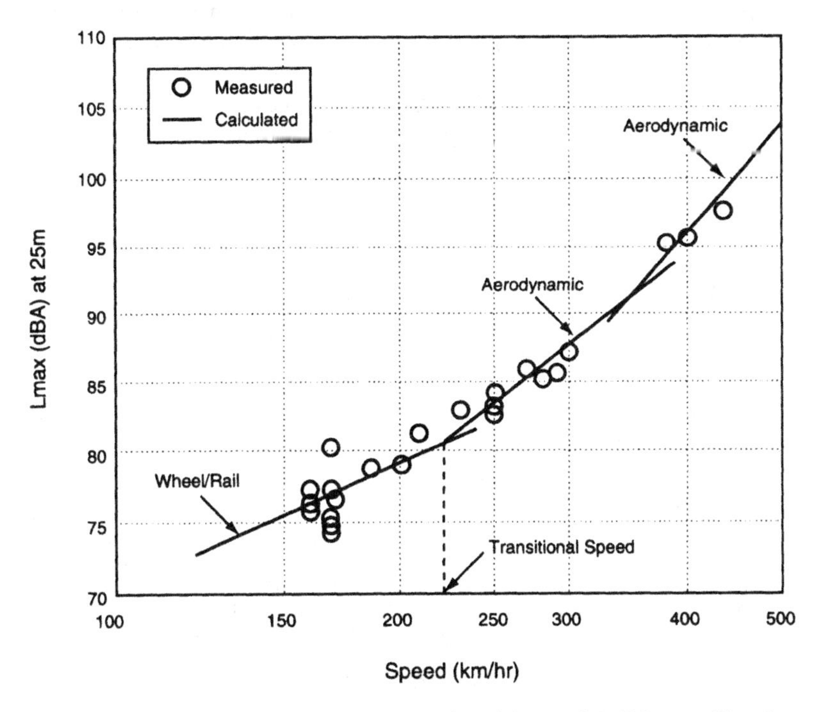

Figure 4.1. Noise vs. Speed Characteristics of a High Speed Rail System Showing Transition Speed.

speed between "wheel/rail noise" and "aerodynamic noise." Note that the level-vs-speed relationship becomes a straight line when plotted on a semi-log graph, thereby making it much easier to interpret noise and speed relationships.

Noise level also increases when wheels and/or rails develop rough running surfaces. For modern high speed trains with wheels and rails in excellent condition and for maglev systems with low mechanical noise, the transition speed occurs approximately 240 km/hr (150 mph). However, when the wheels or rails become rough, the hypothetical wheel/rail noise line plotted in Fig. 4.1 displaces upward with the same slope parallel to the base line. As a result the transition occurs at a higher speed than for a system with smoother wheels and rails. As a result, for electrically-powered trains with rough wheels or rails, the aerodynamic noise sources may not be heard over the roar of the wheels even at very high speeds.

Actual noise levels measured from high speed trains confirm the general principles discussed above, as shown in Figure 4.2. Here, the noise level during a passby is plotted against speed for a variety of train types, including: U.S., European and Japanese electric steel wheel/ steel rail trains (Amtrak AEM7, French TGV, German ICE,

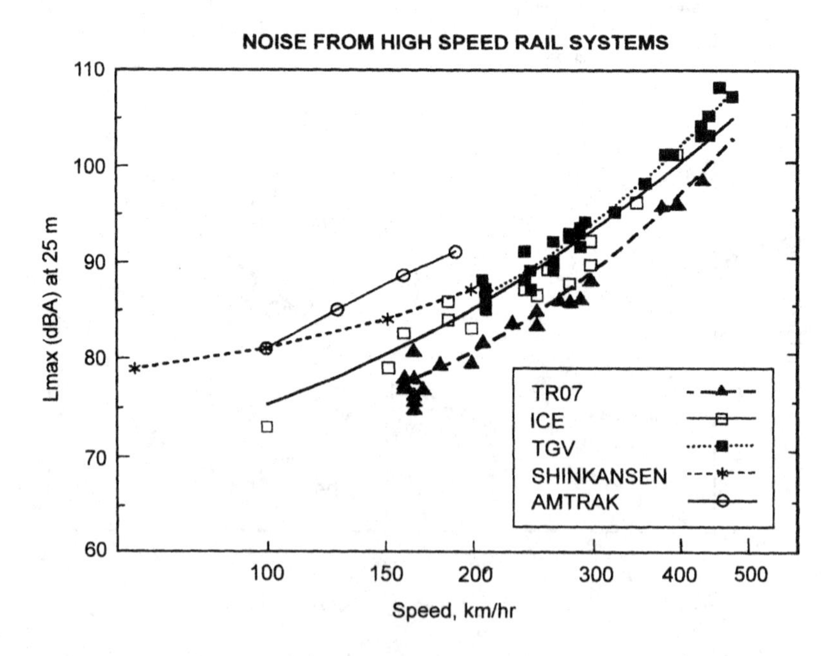

Figure 4.2. Noise Data From High Speed Trains.

Japanese Shinkansen) and a German maglev test train (TransRapid TR07). Noise from these trains show the characteristics of wheel/rail noise at speeds below 200 km/hr (125 mph) and aerodynamic noise at speeds above 250 km/hr (156 mph). The slopes of the curves in those particular speed ranges are consistent with speed to the third power and to the sixth power, respectively. The maglev system exhibits the same type of relationships as do the steel wheel systems, mechanical noise at low speed and aerodynamic noise at high speed, only the levels are generally lower overall. Mechanical noise from a maglev system is generated from vibrations of the car body structure and the guideway structure as the vehicle moves along the guideway. Since the maglev vehicle is rigidly held in place by the magnetic field, dynamic loads are transmitted much as if the vehicle has direct contact. Consequently, deflections or vibrations of the vehicle can transmit to the structure and can be re-radiated as noise.

Noise Sources

Noise from a high speed rail system is generally dominated by three sources: the propulsion and auxiliary equipment, mechanical/structural radiation and airflow moving past the train. The sources differ in where they occur on the system (Figure 4.3) and in what frequency range they dominate.

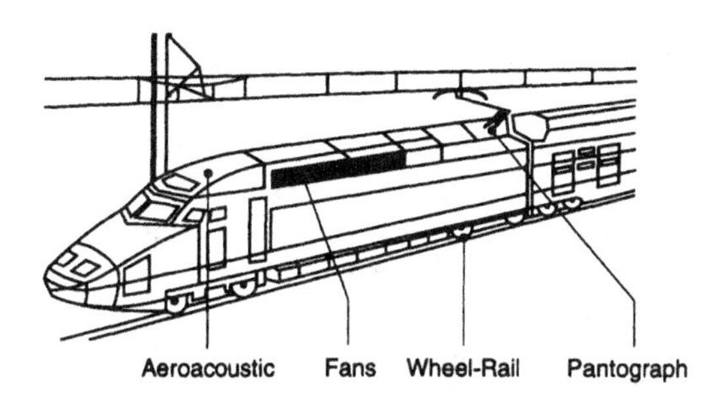

Figure 4.3 Noise Sources on High Speed Train.

Propulsion noise sources

High speed trains are generally electrically powered; the propulsion noise sources are those from electric traction motors or electromagnets, control units and associated cooling fans. Fans have been found to be a major source. On these trains, major cooling fans are located near the top of the power cars, about 3.5 m above the rails, as indicated in Fig. 3.3. Noise from the magnets in a maglev system is a result of induced vibration from magnetic forces. These forces are located at the magnet gaps between the vehicle and the guideway, and sound radiation can come from there as well as from larger structures (vehicle panels, guideway, etc.) caused to vibrate in response to such forces.

Diesel- and gas turbine propulsion systems are not generally considered candidates for high speed operations, although trains hauled by gas turbine locomotives have been shown to be capable of speeds approaching 200 km/hr (120 mph). Fossil fuel propulsion units are usually configured in three parts: a power generation unit, an alternator and electric traction motors which drive the wheels. All three elements generate noise, but the dominate source of noise radiated to the wayside is the exhaust located on top of the locomotive, about 4 m above top of the rails.

Mechanical/structural noise sources

The effects of wheel/rail interaction on conventional trains, guideway structural vibrations, and vehicle body vibrations fall into the category of mechanical noise sources. Wheel/rail interaction is the rolling noise radiated by steel wheels and rails caused by small roughness elements in the running surfaces. This noise source is close to the trackbed with an effective height of about 0.8 m above the rails and generally dominates the A-weighted sound level of conventional trains at low speeds. However, wheel/rail noise can be effectively shielded by low barriers.

Other mechanical noise sources are guideway vibrations and vehicle body vibrations. Both of these sources tend to radiate sounds at very low acoustical frequencies: fundamental resonance frequencies of guideway support beams are generally below 10 Hz, with radiation from box beam panels up to about 80 Hz. Vehicle body vibrations depend on the details of skin and body panel construction, but they can result in significant sound radiation throughout the audible range.

Maglev technology is not free from mechanical/structural sources despite the contactless nature of the system. The maglev analogies to wheel/rail noise from a conventional train are:

1 noise from wheels rolling on guideway support surfaces at low speeds for electrodynamic levitated systems (this type of maglev requires forward motion before lift can occur), and
2 noise from magnetic pole passing.

Moreover, maglev guideway structure is subject to similar loading forces as a conventional train, leading to similar vibrations and radiated sound from the guideway. The vehicle body constructions may also be similar to conventional train cars in response to dynamic forces, resulting in similar vibration and sound radiation characteristics.

Aeroacoustic noise sources

Noise from airflow over a train is generated by flow separation and reattachment at the front, turbulent boundary layer over the entire surface of the train, flow interactions with edges and appendages, and flow interactions between moving and stationary components of the system. These sources can be located over the entire surface of the train and at the edges of guideway structure.

Vibration

In comparison with noise, groundborne vibration is less widespread as an environmental problem. People exposed to vibration tend to be those whose homes or places of work are near a city street with buses or trolleys, or near a highway or railway. When a vehicle passes by, these people may experience feelable movement of the building floors, rattling of windows, shaking of items on shelves or walls, and rumbling sounds, all of which can lead to annoyance when the events happen many times during a day and night.

Vibration problems tend to be very localized and very much dependent on the severity of vibrations caused by the source and the conditions of the ground between the source and receiver. High speed rail systems have the possibility of creating significant ground vibrations. Steel wheels and steel rails have imperfections in their running surfaces – small bumps and discontinuities. A rolling wheel runs into these discontinuities and sets up vibrations of both the rail and the wheel. These vibrations excite the adjacent ground, creating vibration waves in the soil which transmit through the various soil and rock strata to foundations of nearby buildings.

Since the process is very localized, it is difficult for researchers to generalize the vibration characteristics of high speed trains. Vibrations in the ground from passbys of high speed trains are determined by analysing data obtained from sensors placed along the tracks. A number of vibration studies have been sponsored in the United States, primarily related to urban rail transit systems. A major study on high speed rail vibration has been sponsored by the Federal Railroad Administration.[6] Despite the localized nature of the problems, generalized conclusions of studies can be made:

- Smooth, round wheels and smooth, jointless rails minimize vibrations from high speed trains.
- Maglev trains generate negligible ground vibrations.
- Ground vibrations from high speed trains are rarely noticeable beyond 100 meters from the track.
- Bedrock within 10 meters of the surface can cause vibrations to propagate to greater distances than normal.
- Some soils, such as saturated clay, are particularly conducive to vibrations; others, like dry sand, damp out vibrations.
- Buildings made of light weight materials tend to respond more greatly to ground vibration waves than do heavy, large buildings.

4.1.2. Criteria for Noise and Vibration Impact Assessment

Noise Criteria. Reaction to noise in the environment varies with the individual. Some people can sleep in a noisy urban setting, and some people are awakened by the slightest sound. Because of the wide range of variablility in public reactions to noise, it is necessary to focus on community- wide average reactions obtained from large scale attitudinal surveys in order to develop general criteria for noise impact. Establishment of noise levels for impact criteria is not intended to eliminate all potential for annoyance from the noise, but to identify areas where communities on the average would find the noise to be unacceptable and where noise mitigation should be considered. Designers of high speed rail projects must ensure that all reasonable steps are taken to keep train noise levels from being an unreasonable burden on residents and other noise- sensitve receivers.

Characterizing the noise environment in the vicinity of a high speed ground transportation system involves more than describing how much noise is made by a single train passby. For example, trains pass throughout day and night, but some periods are more noise-sensitive than others; trains come in various lengths, but longer trains make noise over a greater duration of time; main lines carry more trains than branch lines, but more noise events have greater impact than few events. For the foregoing reasons, the cumulative noise measures such as the Leq and Ldn are appropriate for assessing the total contribution of trains to the noise environment. Acoustical scientists have devoted much effort to relating these measures of noise exposure to public reaction to come up with criteria for noise impact.

Social surveys have been conducted throughout the world to form bases for noise criteria. The U.S. Environmental Protection Agency focused on annoyance and health effects of noise in extensive research conducted during the 1970's. This research provided the basis for the development of noise descriptors and criteria by other federal agencies in the United States. A key result was a synthesis of many attitudinal surveys to develop a universal curve relating annoyance to measured noise levels, expressed in terms of Ldn. This curve, recently updated, is used as the basis for developing noise criteria from transportaion sources in the United States.[5] Therefore, the descriptor introduced earlier in this chapter, the Ldn, is the one that satisfies two key requirements: (1) it accommodates all noise sources with various magnitudes, durations and times of day, and (2) it has been found to correlate well with the results of attitudinal surveys of residential noise impact.

The United States is currently developing new noise criteria for assessing environmental impact of high speed rail systems.[6] The criteria are consistent with those developed for the maglev concepts during the National Maglev Initiative.[2] The criteria include a comparison of project noise with existing noise expressed in terms of Ldn, but with an added consideration of startle effects due to rapid onset rates of high speed rail noise signatures. Startle effects associated with rapid onset rates of noise events have been investigated in research sponsored by the U.S. Air Force.[7] Since the noise signatures caused by high speed rail and maglev trains close to the tracks are similar to those that are known to cause startle from aircraft, there is reason to believe that startle may be a factor in annoyance from these sources also. The proposed criteria are shown in Fig. 4.4. Noise criteria are defined for various land use categories depending on their sensitivity to noise. These include: Category 1: Lands particularly sensitive to noise; Category 2: Residential areas and buildings where people sleep; Category 3: Institutional land use where the noise sensitivity occurs primarily during daytime.

Other countries with high speed rail systems have proposed or have adopted noise criteria for environmental assessment purposes. These are summarized below:

Japan – Japan was the first country to establish environmental noise limits for high speed trains. As a result of expressed public

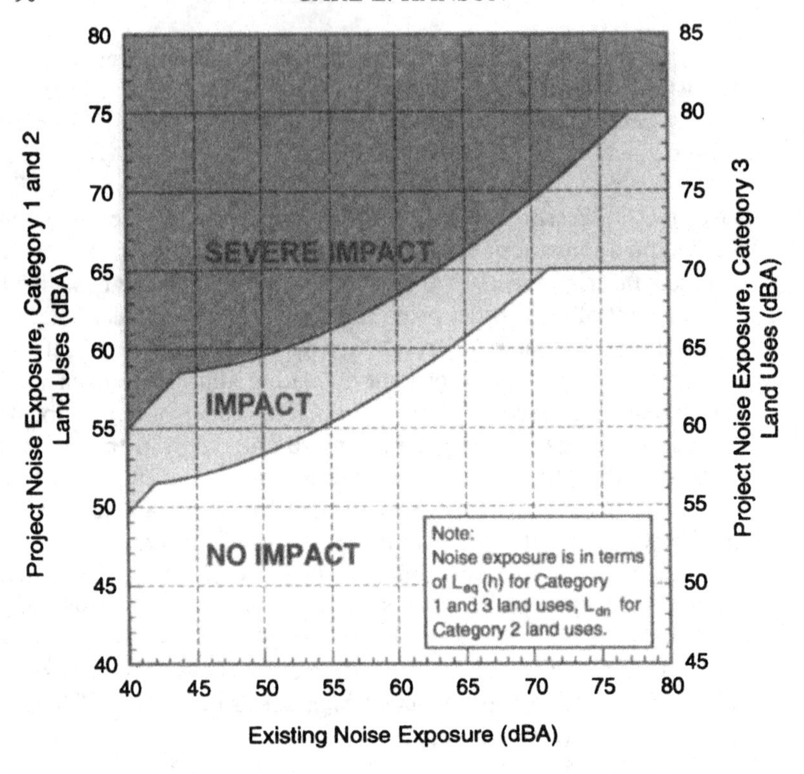

Figure 4.4. Proposed Noise Criteria for U.S. High Speed Rail Projects.

concern, the authorities established maximum noise levels for train passbys of Lmax = 75 dBA at a reference distance of 25 meters (82 feet). No provision is made for penalties of night noise events or for numbers of passbys. However, since the Shinkansen service is nearly continuous, with only a few minutes between train passbys, and because the limit is applied to trains only and not for comparison with other noise sources in the environment, the use of a maximum limit may be appropriate in this case.

France – France uses the daytime Leq applied to train events that occur between the hours of 8 a.m. and 8 p.m., with a mitigation level established at Leq = 65 dBA.

Sweden – Sweden has no official guidelines, although recommendations have been made to consider mitigation for residential buildings where the maximum level from trains exceeds 80 dBA and where the 24-hour Leq exceeds 55 dBA. In locations where there is "strong vibrations" from trains in addition to the noise, the limits are proposed to be 10 dBA more stringent.

Vibration Criteria In contrast with noise, community reaction to vibration levels from high speed trains has not been thoroughly researched. It is known that people are not concerned about outdoor vibrations, but become annoyed when vibration levels inside buildings caused by an external source exceed a limit, even if self-generated vibrations are shown to be greater. International and national standards have been proposed for environmental impact, and lacking further verification, these are generally adopted. The U.S. Department of Transportation is currently developing vibration criteria for assessing environmental impact of high speed rail systems.[6] Generally, impact is considered to occur for vibration levels exceeding a level of Lv = 72 VdB for frequent events (more than 70 events a day), or Lv = 80 VdB for infrequent events. Special criteria have been established for certain sensitive industrial, commercial and research activities, and for avoidance of damage from some construction activities. Similar criteria are used by the U.S. Department of Transportation, Federal Transit Administration for urban transportation systems.[10]

4.1.3. Method for Assessing Noise and Vibration Impact

Noise assessment for a new high speed rail project is performed in three stages as the project develops. In the early planning stage a **Screening Procedure** is used to determine whether any noise-sensitive receivers are within a distance where impact is likely to occur. Where there is potential for noise impact, the next two levels of analysis are employed, **General Assessment** and **Detailed Analysis**. A general assessment can provide the appropriate level of detail about noise impacts for a "corridor" study which is undertaken during the planning of a high speed rail project. The procedure involves noise predictions commensurate with the level of detail of data available in the early stages of major projects. A **Detailed Analysis** is undertaken when the greatest accuracy is needed to assess impacts and the effectiveness of mitigation measures on a site-specific basis. In order to do this, the project must be defined to the extent that location, alignment, mode and operating characteristics are determined. The results of the Detailed Analysis would be used in predicting the effectiveness of noise mitigation measures on particular noise-sensitive receivers.

This section provides information for the application of a Screening Procedure. For procedures related to General Assessment and Detailed Analysis, the U.S. Department of Transportation is preparing methods to be used.[6] Another source of analysis procedures is the guideline published by the same agency for urban transit system noise and vibration assessments.[10]

Screening Procedure. The purpose of the noise impact screening procedure for a high speed rail project is to define an area within which the project may have an adverse effect for the noise-sensitive land uses listed in Section 4.1.3. The distances given in Tab. 4.1 delineate a project's noise study area. The areas defined by the screening distances are sufficiently large to encompass all potentially impacted locations. They were determined using estimates that were conservative in terms of operating characteristics and source levels, in order to determine the greatest possible impact area so as not to leave out any potentially affected locations.

The first step in the procedure is to select the type of project from the candidates listed in Tab. 4.1. The three types of projects represent a generalization of many high speed ground transportation systems and may not exactly fit individual projects. All the systems are assumed to be electrically powered. These are summarized below:

TABLE 4.1. Screening Distances for High Speed Rail Noise Assessments

TYPE OF PROJECT	*SCREENING DISTANCE (meters)*	
	UNOBSTRUCTED	*OBSTRUCTED*
High Speed Rail	100	50
Very High Speed Rail	250	125
Ultra High Speed Ground Transportation (Maglev)	200	100

High Speed Rail: This category includes the conventional inter-city steel wheel/steel rail system with maximum speeds of about 200 km/hr (125 mph). Typical examples of trains in this category in 1995 are the Swedish X2000 and Amtrak's Metroliners on the Northeast Corridor between Washington, D.C. and New York.

Very High Speed Rail: This category includes the high speed steel wheel/steel rail systems in operation in Europe and Japan with typical operating speeds of 300 km/hr (188 mph). Representative systems are the Japanese Shinkansen, the French TGV and the German ICE trains, all in everyday operation.

Ultra High Speed Ground Transportation: Only one type of system is likely to be able to reach the ultra high speeds characteristic of this category – magnetically- levitated (maglev) systems with maximum speeds of 500 km/hr (312 mph). Maglev test trains in Japan (Linear Express) and Germany (TransRapid) have demonstrated feasibility of the technology but it remains to be put into revenue service.

Key parameters in any noise analysis are the schedule and the length of trains, information that results from the patronage analysis for each corridor. Assumptions are often made at the earliest stages of environmental assessment where the screening procedure will be used. Information in Tab. 4.1 assumes 24 trains per day with an average length of 200 meters.

The second step is to review the alignment on the corridor maps, preferrably showing land use and terrain features. If the alignment is known, then the distances shown in Table 4.1 for the selected system of interest can be marked on the maps. Unobstructed distances refer to open land with no intervening rows of buildings or obvious terrain features that would provide a screen for the train. Any noise-sensitive land uses that lie within the marks should be inventoried as sites potentially impacted by noise.

Clusters of such potentially impacted sites will be candidates for noise abatement. A more detailed analysis is called for in these cases. More detailed methods of analysis are being developed by Federal Railroad Administration.[6] Examples of similar methods have been published by the Federal Transit Administration.[10]

Vibration Screening Procedure

Vibration problems associated with high speed rail systems tend to be localized and site-specific depending on the soil conditions between the tracks and the vibration-sensitive receiver. A key factor appears to be the depth to bedrock. When rock strata are within 10 meters of the surface, ground-borne vibrations tend to propagate efficiently, resulting in vibration problems at greater distances from the track than for less stratified conditions. Other types of soil conditions have varying vibration propagation characteristics as well, making it difficult to generalize. For example, dry sand tends to damp out vibrations, whereas saturated clay soil tends to transmit vibrations efficiently.

The information necessary to estimate vibration propagation characteristics of the soil on a site-specific basis is not generally available at the stage of a project when a screening procedure would be used. Consequently, the initial screening is based on two different soil conditions, "Efficient" and "Normal." The distances within which vibration impact could occur are given in Table 4-2. Note the Ultra High Speed System is assumed to be maglev which transmits insignificant vibration to the ground.

TABLE 4.2 Screening Distances for High Speed Rail Vibration Assessments

TYPE OF PROJECT	*SCREENING DISTANCE (meters)*	
	NORMAL	*EFFICIENT*
High Speed Rail	35	70
Very High Speed Rail	40	80
Ultra High Speed Ground Transportation (Maglev)	0	0

4.1.4 Mitigation Methods for High Speed Rail/Maglev Systems

Noise Abatement: Noise transmitted from high speed ground transportation sources can be controlled at three places: by quieting the source of the noise, by interfering with the transmission path from source to receiver, or by treating the buildings receiving the noise. Controlling noise at the source is the most effective method of abatement, since the treatment reduces noise all along the alignment. The other treatments are site-specific treatments and affect a limited area or single buildings. There are many noise sources associated with a high speed train, as described in Section 4.1.1. Builders of high speed trains and maglev systems expend significant effort to locate and diagnose each source and to take countermeasures to abate their noise levels. A major noise reduction program in Japan has been undertaken to combat noise on Shinkansen.[4] Following the Japanese lead, other countries undertook research program to reduce noise from high speed trains; notable references are the German program to reduce noise from the ICE and to control noise from the maglev vehicles planned for use in the Berlin-Hamburg Corridor.[3]

Examples of noise abatement methods for high speed trains include:

Source controls—wheel dampers applied to solid steel train wheels to reduce wheel noise; skirts over wheels to block radiated sound; smooth body shapes and pantograph modifications to reduce aerodynamic sound at high speed;

Path control—sound barrier wall between the source and the receiver; and,

Receiver controls—sound insulation of walls, doors, and windows of buildings facing the tracks.

Vibration Abatement: Similar to noise abatement methods, control of vibration involves the same three elements: control of vibrations at the source, along the transmission path and at the receiving building. However, in contrast with noise control, vibration abatement methods are not as well developed. This is due to the extremely complex nature of the generation and propagation of vibrations and the difficulty in control. For example, interfering with the path of vibration waves in the ground is made difficult because of their long wavelengths, typically 10 meters or more. As a result, most vibration control efforts are directed toward the source and its connection to the ground in order to minimize the generation of vibrations at the initial stages of the process.

Examples of vibration abatement methods for high speed trains include:

Source controls—wheel/rail smoothing for running surfaces of solid steel train wheels; elimination of discontinuities such as joints in rails; special track support systems such as resilient rail fasteners;

Path control—rubber or foam mats beneath ballasted track; trenches between the source and the receiver; and,

Receiver controls—resilient mounts in the foundations of buildings; isolation of foundations with resilient layers.

4.2 ELECTRIC AND MAGNETIC FIELDS OF HIGH SPEED RAIL/MAGLEV SYSTEMS

4.2.1 Technical background: definitions and characteristics

Background

There is growing public concern about possible health effects and interference with communications due to exposure to low frequency electric and magnetic fields (EMF). Much community apprenhension tends to be focused on high-voltage electric transmission lines, but the public does not generally discern the difference between transmission lines and catenary lines for electric trains. This section summarizes the findings of recent research into the effects of EMF on humans as it relates to the kinds of fields experienced in the vicinity of high speed rail and maglev rights-of-way.[11]

Definitions

Electric power is supplied to high speed trains from nearby substations near the tracks to the train via overhead catenaries (suspended wires) or "third rails" running alongside the guidance rails. Electric current returns through the rails back to complete the circuit. Whenever current flows through a loop an electromagnetic field (EMF) is formed within the loop and in a region around it. The magnitude of the magnetic field is proportional to the magnitude of the current in the loop; it oscillates corresponding to the frequency of the electrical current. At very high intensities EMF hazards include electric shock, pacemaker interference and burns, but the magnetic fields in the vicinity of transportation systems are classified as low intensity fields where the effects are not well known. Since the frequencies used to transmit current are 0 Hz (Direct Current), 16.67 Hz (some parts of Europe), 25 Hz (Amtrak south of New York), 50 Hz (most of Europe), or 60 Hz (most of North America), the magnetic fields near transportation systems oscillate at what is termed extremely low frequencies (ELF).

Concern is focused primarily on magnetic fields rather than electric fields. Unlike electric fields, magnetic fields are difficult to shield; consequently they encompass our environment without an apparent means of reduction. They also penetrate biological tissue (our bodies) quite easily whereas electrical fields concentrate on skin surfaces.

The commonly used unit of magnetic field intensity (flux density = concentration of magnetic field)is either "Gauss" or "Tesla" (1 Tesla = 10,000 Gauss). This section expresses magnetic intensity in Gauss because most human exposure is in fractions of that unit.

Common Exposures to ELF EMF

Humans are exposed to ELF EMF to varying degrees in everyday life. The universal exposure is the earth's geomagnetic field which ranges from 300 to 800 milliGauss, averaging out to 500 milliGauss (0.5 Gauss). During our everyday lives, we are exposed to magnetic fields near electric appliances ranging from about 1 to 1000 milliGauss (.001 to 1 Gauss). Workers in the electrical service field can be exposed to fields of 10 to 100,000 milliGauss(.01 to 100 Gauss).[12] The major difference among the exposures is the varying frequencies – the earth's field is steady state (0 Hz) whereas most of the rest of our exposure is related to the line frequency of our local electric current (60 Hz, in U.S.).

A number of epidemiologic studies have taken place to determine the increased risk of cancer and other diseases for people exposed to ELF EMF.[13] Although a number of problems have been reported in the literature, the results have not in general been verified in studies carried out in western nations. In summary, no clear pattern has emerged although research is continuing.

4.2.2. Characteristics of EMF Associated with Transportation Systems

Research on EMF in the vicinity of transportation systems has recently been undertaken by the Federal Railroad Administration in anticipation of the introduction of new high speed electrically-powered trains into the U.S.[4] Measurements of magnetic fields were taken in a wide variety of situations – inside passenger compartments, at operator's positions, on platforms and at the wayside near power lines. As might be expected, the variation in exposures inside vehicles was significant, ranging from a few milliGauss to 2.5 Gauss, depending on the design of the vehicle's electrical distribution system. Outdoors, however, the static fields differed very little from the earth's magnetic field of 500 milliGauss. Oscillating field strength at line frequency falls off with distance away from the track in a fairly predictable manner, as shown in Fig. 4.5.

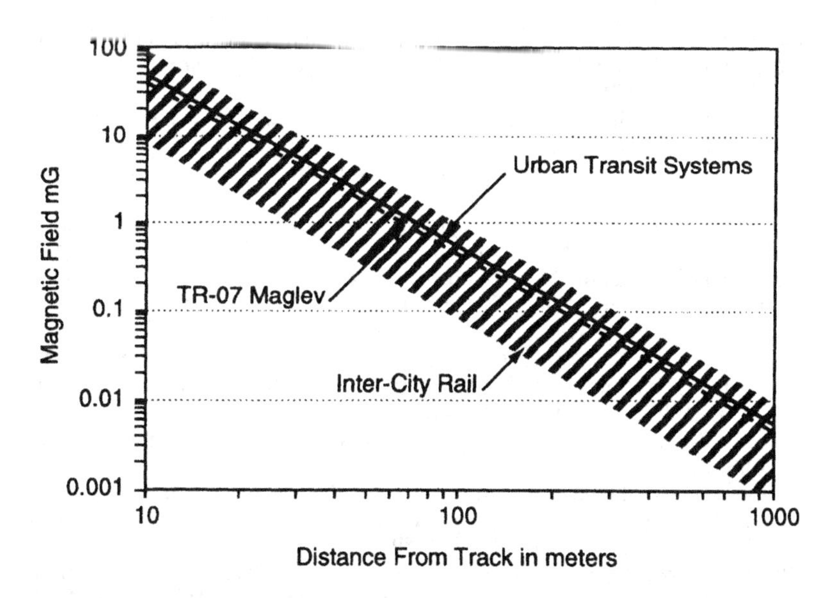

Figure 4.5 Magnetic Field Attenuation as a function of Horizontal Distance from the tracks (From Ref. 4).

4.2.3. Criteria for EMF Impact

There are no federal standards in the U.S. to establish limits on the exposure to ELF EMF. Two states , Florida and New York, have adopted standards to limit EMF at the boundaries of transmission line rights-of-way.[5,6] Other organizations have established guidelines, including the World Health Organization,[7] the International Radiation Protection Association,[8] and the American Conference of Governmental Industrial Hygienists.[9] In general, the measurements conducted by U.S. Department of Transportation indicate that ELF EMF's in and around electrical transportation vehicles are at least an order of magnitude less than the strictest of the exposure guidelines in use today.[4]

NOTES

1 Harris, Cyril M. (ed.) *Handbook of Acoustical Measurements and Noise Control*, Third Edition, McGraw-Hill, New York, 1991.
2 Hanson, C.E., Philip Abbot, Ira Dyer *Noise from High Speed Maglev System*, U.S. Department of Transportation, Federal Railroad Administration, Report Number DOT/FRA/NMI–92/18, October 1992.
3 King, W.F. "The Components of Wayside Noise Generated by High-Speed Tracked Vehicles," *Proceedings InterNoise 90*, Gothenburg, Sweden, August 1990.
4 Zenda Y., et al. "Noise Control of High Speed Shinkansen," Proceeding of the Fifth International Workshop on Railway and Tracked Transit System Noise, Voss, Norway, 21–24 June 1995 (publication in *Journal of Sound and Vibration*, May 1996)
5 Fidell, S., D.S. Barber, and T.J. Schultz. "Updating a Dosage-Effect Relationship for the Prevalence of Annoyance Due to General Transportation Noise," Journal of Acoustical Society of America, Vol. 89, No. 1, January 1991.
6 Hanson, C.E., et al. *High Speed Ground Transportation Noise and Vibration Impact Assessment*, U.S. Department of Transportation, Federal Railroad Administration, Report to be published December 1997.
7 Stusnick, E., et al. "The Effect of Onset Rate on Aircraft Noise Annoyance." US Air Force Report AL/OE-TR-1993–0170, October 1992.
8 Öhrström, E., A. Skånberg. "A Field Survey on Effects of Exposure to Noise and Vibrations from Railway Traffic." Proceedings of Fifth International Workshop on Railway and Tracked Transit System Noise, Voss, Norway, 21–24 June 1995 (publication in Journal of Sound and Vibration, May 1996).
9 International Standards Organization. "Evaluation of Human Exposure to Whole-Body Vibration, Part 2; Continuous and Shock-Induced Vibrations in Buildings (1–80 Hz)," ISO–2361–2, 1989.
10 U.S. Department of Transportation, Federal Transit Administration. *Transit Noise and Vibration Impact Assessment*. Report DOT-T-95–16, April 1995.

11 U.S. Department of Transportation, Federal Railroad Administration, Office of Research and Development. Series of Reports on Safety of High Speed Guided Ground Transportation Systems, August 1993.
12 Goellner, Don, et al. "Safety of High Speed Guided Ground Transportation Systems – Broadband Magnetic Fields: Their Possible Role in EMF-Associated Bioeffects," U.S. Department of Transportation, Federal Railroad Administration, Report No. DOT/FRA/ORD-93/29, August 1993.
13 Creasey, William, and Robert B. Goldberg. "Safety of High Speed Guided Ground Transportation Systems; Potential Health Effects of Low Frequency Electromagnetic Fields Due to Maglev and Other Electric Rail Systems," U.S. Department of Transportation, Federal Railroad Administration, Report No. DOT/FRA/ORD – 93/31, August 1993.
14 Dietrich, Fred M., William Feero and William Jacobs. "Safety of High Speed Guided Ground Transportation Systems: Comparison of Magntic and Electric Fields of Conventional and Advanced Electrified Transportation Systems." U.S. Department of Transportation, Federal Railroad Administration, Report No. DOT/FRA/ORD-93/07, August 1993.
15 Florida Administrative Code, Chapter 17–274. "Electric and Magnetic Fields."
16 State of New York Public Service Commission. "Interim Policy on Magnetic Fields of Major Electric Transmission Facilities." September 11, 1990.
17 World Health Organization. "Environmental Health Criteria 69: Magnetic Fields." Geneva, 1987.
18 International Radiation Protection Association "Interim Guidelines on Limits of Exposure to 50/60 Hz Electric and Magnetic Fields." *Health Physics* 58: 13–22. 1990.
19 American Conference of Governmental Industrial Hygienists. "1992–1993 Threshold Limit Values for Chemical Substances and Pysical Agents and Bilogical Exposure Indices." Second Printing.

CARL E. HANSON

Dr. Carl E. Hanson is a Senior Vice President and an original founder of Harris Miller Miller & Hanson Inc. located in the Boston area, one of the premier acoustical consulting firms in the United States. He is recognized as this country's leading authority on noise and vibration from high speed rail and maglev systems. Among his major contributions to the field have been the development of the methods for assessment of noise and vibration impact from both high speed rail and maglev for the Federal Railroad Administration, as well as advising Amtrak on the noise and vibration specifications for high speed trainsets. In his twenty five years as a transportation noise specialist, he has worked for railroads, transit authorities and government agencies throughout the United States and has represented this country at rail noise workshops and conferences throughout the world.

Dr. Hanson earned a BS in Aero Engineering at University of Minnesota, a M.S. in Mechanical Engineering and a Ph.D. in Acoustics at Massachusetts Institute of Technology. He is a Registered Professional Engineer in Massachusetts, Connecticut, Florida and Texas.

5 FINANCING HIGH SPEED RAIL INVESTMENTS – THE VIEW FROM EUROPE – LESSONS FOR THE UNITED STATES – THE GROWING ROLE OF THE PRIVATE SECTOR

Dr. Tim Lynch,
Director
Center for Economic Forecasting
and Analysis
Institute for Science and Public Affairs
Florida State University
Tallahassee, Florida, 32306

INTRODUCTION

Western, Central and Eastern Europe are in the midst of one of the greatest economic transitions in their collective history. The major shifts in the swiftly unifying European continent include responding to:

- the economic and social integration of Western European economies and cultures and resulting disappearance of "artificial" economic trade borders and barriers;
- the recent crumbling of Communism;
- the opening of Central and Eastern Europe;
- the opening of Russian, Asian and Middle-eastern markets;
- the imploding globalization of the world economy and the fierce international competitiveness of the booming U.S., Canadian and Mexican North American Free Trade Agreement (NAFTA) economies;
- expansion of the economies of the Asian tigers led by Japan;

- international expansion of world markets and competition driven by the General Agreement on Tariffs and Trade (GATT);
- recent demographic shifts including the aging of European populations; and
- the emergence and growing dominance of service-sector driven high technology markets for the distribution of goods and services. Two high tech advances include the deployment of advanced high speed rail (HSR) systems across the continent and high speed communication technologies across the world.

All of these factors are contributing to the unification of the continent by playing into a dynamic socio-political revaluation of conventional balance of political and economic powers. These shifts are also influential in the revaluation of how the Europeans finance HSR and other infrastructure. This section will review some of the shifts at hand and present a framework of future needs and how the European's HSR financial analysis process evaluates development of infrastructure investments.

5.1. THE EUROPEAN VIEW OF FINANCING INFRASTRUCTURE INVESTMENTS IN THE NEW EUROPE

The financing of HSR in Europe evolved during the past decade from a purely public sector enterprise, to one with increasing levels of support and long-term commitment from the private sector. While national companies and direct public sector investment remain the primary instruments of HSR and other related transportation investment, private sector responsibilities are becoming more well defined. The Director General for the Lending European Investment Bank states:

> In this post-ideological era, after the collapse of communism and the triumph of western civilization and free enterprise, there is one rearguard battle fought between interventionists and planners on one side and the already victorious advocates of free markets, and lesser-faire on the other.
>
> The battle is about the role of government in infrastructure investments. (In England and elsewhere) *the incorrect impression is that this duel will be won by the protagonists of the private sector.* I say incorrect because in most countries of continental Western Europe it is accepted and deemed necessary that the public

> sector keeps on playing a major role . . . . The impression one gets is that the triumph of capitalism over state-run, centrally-planned communism is only definitive when all infrastructure will be privatized.

General Treumann identified three interrelated points drawn from Europe's economic reality:

1. The European Economies are facing serious investment shortfall difficulties in financing new infrastructure;
2. The need for major infrastructure investment will require *governments and the private sector to complement* each others limited resources; and
3. New forms of investment co-operation are needed such as the European Investment Fund currently under development.

The new Cohesion Fund, and other Community funds used for joint European investment initiatives were created to help the emerging peripheral countries (Spain, Greece etc.) and to assist with the development of Eastern Europe. Even with this there is insufficient investment in Europe for needed infrastructure expansion. The solution to resolve this problem, to quote Director Treumann is "to change how the private and public sectors split their responsibility for investment in infrastructure in a co-operative way to co finance these investments."

THE COHESION FUND-FINANCING TRANSPORTATION COMMUNICATION AND ENERGY INFRASTRUCTURE

The absence of an integrated (but not nationally based) infrastructure plan for all of Europe precludes the Single Market from collectively benefiting from macro-investments. The trans-European HSR alignments are one of the primary beneficiaries of the Cohesion fund established by the Maastricht summit, to finance transport infrastructure in the less developed EC countries. This fund was established for the promotion of trans- European networks of transport, such as the International High Speed Rail plan, telecommunications and energy infrastructure.

To achieve these ends and secure associated massive capital infusions, greater amounts of private sector monies must be mobilized. Increasingly, the national governments of Europe will be responsible for:

- creating clear and sound economic conditions to ensure the attractiveness of private capital investments;
- creating sound institutional and planning conditions that will ensure the success of these investments; and
- providing initial capital investments to ensure the financial soundness of these activities to the public and private sectors.

Additionally, users of the infrastructure such as HSR riders will increasingly be required to pay directly for the increased costs of the services that these infrastructure provide. In a typically European view of financing infrastructure, it is recognized by financial institutions across the EC, that ridership fees and other user charges on HSR and related infrastructure will not be sufficient to finance the full cost of development of the HSR systems. Director General Treumann described the central European tenant underpinning the development of European HSR and other transit investments:

> A transport service (unlike CD players, telephone services, energy supply and restaurants, for instance) cannot in most cases be supplied on a profitable or even cost covering basis. Conventional capitalistic wisdom would dictate that the HSR or transport system should be liquidated or phased out as happens with restaurants and manufacturers of CD players if they cannot break even. Only those services and goods are produced which can cover their costs. The benefits perceived by the consumer is least equal to the cost of producing those products.
>
> However (in the European view) most transport services, especially those which need to modernize due to outdated infrastructure, rolling stock or those that need to expand, show huge losses. In some rare cases revenue is barely enough just to cover operating costs. (Sometimes achieved after some creative cost allocations, for which authorities are quite famous, is applied).

PROVISION OF TRANSPARTATION SERVICES YIELD SOCIAL BENEFITS

Why then does society continue to fund these transport services? The answer is clear, the services provided by the HSR and other public transport systems provide important economic and public service benefits that exceed the costs levied for transporting individuals, goods and services. These positive effects on the economy captured by non-using individuals are called externality benefits. They include;

- a cleaner and aesthetically valuable environment,
- time saving and
- energy savings over use of other modes,
- improved accessibility across our urban centers;
- increased economic activity and;
- generation of employment that exceed the increases from similar expansions of other modes of transport.

To illustrate, sophisticated computer models estimate inter-urban travel across Europe[2] and evaluate HSR and other transportation mode ridership forecasts for the year 2000 and 2010. These models predict that investment in moderate or basic HSR system will result in an increase in overall European population movements between 1.85 and 2.2% by 2010. The deployment of the full HSR train network would result in a 3% increase in interzonal mobility of European passengers across the continent. This reflects an expansion of the European economy with concomitant increases in flows of goods and services that otherwise would not occur without this investment in HSR. This, in it's simplest form is an example of net economic expansion and diversity across the European markets generated exclusively by HSR investments.

These benefits to non-system users result in transfers of economic value to private individuals from investments typically financed by the public sector that exceed the revenues collected by the HSR and other transport services. The benefits of a cleaner environment or increased economic activity accrue to private individuals or across the general economy (i.e., to all citizens). These nonuser beneficiaries are said to experience the "free rider effect", they can benefit without paying unless assigned a user fee or tax. Economic theory suggests that the prime source of financing these benefits should rest with the general (national) economy and its citizens through publicly-funded taxes. Director General Treumann emphasizes this point:

> There is only one institution that can implement such transfers (back to the public sector from the general economy), and that is a government, be it local or national. The contribution of the users can be coaxed out of them in exchange for the ticket, for the non-users who benefit anyway whether they pay or not, enforcement, that is to say taxation to get their contributions, is essential.
>
> The basic thrust of this argument is that irrespective of what mixture of public and private HSR (or transit) financing and ownership is proposed there is always a need for a basic agreement between all participants on the desired level of service to be pro-

vided. Operation of a HSR or transit system must also include an explicit agreement and willingness to accept a public subsidy for the system to be financed out of taxes or user fees to capture the value of the free rider effect.

5.2. FUTURE CAPITAL REQUIREMENTS AND THE RESULTING NEED FOR NEW PUBLIC- PRIVATE RELATIONSHIPS FOR FINANCING INFRASTRUCTURE ACROSS EUROPE

Recent data indicate total infrastructure investments in transport have ranged from between *60 to 70 billion ECU per year* across the European Community. Future forecasts predict new demands over the next 15 years exceeding 100 billion ECU of which 80 billion will be for transport. The highest priorities include:

- creating a European network of high-speed trains to carry passengers and goods within the framework of an overall policy boosting rail for environmental reasons
- reducing air and roadway congestion with their enormous costs;
- investing in airport enlargement and improvement of connections between airports and major urban centers (with HSR and other modes); and
- developing intermodal facilities with efficient infrastructure linking capacity among the various transport modes.

These investment demands will command 20 billion ECU per year, or approximately 20% of the forecast aggregate European transport investment requirements. During the 1960s, 1970s and 1980s almost all huge European infrastructure investments were financed exclusively by the public sector. The European economies of the 1990s and beyond face a very different setting, beset by substantial additional demands on public budgets for health care, education, welfare, plus the new needs of financing the well-being of Eastern Europe investments. These needs combined with high public sector debt, slower rates of growth and savings and generally higher real interest rates result in insufficient public revenues to finance all future European HSR and other infrastructure needs.

The realization of large-scale infrastructure projects is traditionally guaranteed in Europe, except for sporadic cases, by the financial intervention of the public sector both at national and local levels. There is a new widespread awareness across Europe that the private sector will be increasingly essential in financing and help achieving a more efficient operation of a variety of infrastructure investments across Europe.

These new public-private relationships will be increasingly important since the public sector is responsible for managing all of the permitting and regulatory settings within which all of these projects are planned, designed and implemented.

5.3. MODELS FOR ORGANIZATION OF THE PUBLIC-PRIVATE RELATIONSHIP

One model for this new public-private organization conceptualizes a corporation as a private company whose shares are held by the government. This model allows the company to borrow capital with government backing at very competitive rates.

A second model is a company established with mixed private and public ownership. This is a model that is often used in France with considerable success.[3] This model allows each sector partner to complete respective infrastructure tasks in an efficient manner. Efficient private management extends from the private side while efficient management of government public policy (environmental permitting, review etc.) and preferred interest rates advance from the government side. These realigned priorities increasingly emerge as essential to the successful development of any plan proposed in the United States.

A sound performance and investment contract is needed if the private sector investors are to be attracted and retained within this partnership. As happened with the Eurotunnel and elsewhere across Europe private sector large scale construction firms are seduced into this form of partnership as a point of leverage to capture a considerable profit on the large infrastructure construction contract.

Sometimes this model may make the banking industry somewhat uneasy where they become squeezed between the large share holding construction company as both management and performing contractor on a project with government as a vested interest partner. But this form of private-public partnership is advantageous for transportation projects as it sets up an integrated management team. This team provides clear communications on matters of finance, and efficient private management principals to the public shareholders, and efficient permitting and planning functions and linkages to low interest loans and public sector backing. As the European Investment Bank Director General stated:

> I'm not very positive about the 'pure' private firm, mainly because the involvement of the (public) authorities is too low. Public transport is very integrated in the basic functions of government so that it is very difficult to invest and run a transport system without constant involvement by the authorities. In case of problems, cost overruns, extra environmental or safety requirements, and integration with other transport systems, *the coordinating presence of government (as a full fledged partner in the project with vested financial and other interests) is a must . . . sitting next to the promoter rather than opposite*. (emphasis added)[4]

Summary

In summary, massive future HSR infrastructure investment needs in Europe will require, often for the first time, vast infusions of private capital. Secondly, these new investments will require new and expanded joint public-private partnerships with each partner playing an increasingly important role: the public sector as the infrastructure funding foundation, public policy guardian and facilitator, and the private sector as the efficient manager and new source of capital with increasingly longer (e.g., 20 year plus) return horizons.

One potentially feasible option would involve the public sector covering much of the rail's short run operational losses (covering the hole) and thereby, allowing the private sector partners the ability to earn shorter run returns on the construction of the facilities, and some initial operational profits. These two streams of earnings could allow the private sector to retain an ongoing interest, and await the higher long run profits from operation (above operation costs only). This combination of profitable earning streams may make these projects attractive to the large scale long range visionary contractor-operator, as well as by the private sector.

HSR PRIVATE-PUBLIC PARTNERSHIP IN THE U.S.

In the U.S. fledgling efforts focusing on similar private-public alignments are evolving but with less spectacular results, to date, on HSR project development. In every major corridor within the U.S. where HSR was proposed, from California, Ohio and Pennsylvania to Nevada, Texas, Florida and New York, a number of the nation's largest construction firms were very active as team members in proposed public/private partnership consortiums.[5] To date, however, this membership has not produced conclusive leadership that resulted in a single successful project. The leadership exception is the 1997 cooperative private public relationship forged between the Florida Overland Express (FOX) and Florida DOT described in the last chapter of this book.

To date, the strengths of the partnerships in the U.S. are not as cohesive or inclusive as those formed in Europe. There is less of a history of working together and less of an established trust within these relationships in the U.S. than in Europe. In other sectors of construction and operation, such as airport, shipping and waterway transportation and operation, these private-public partnerships are far more successful. In these settings with the public sector serves as the foundation of funding and support for the infrastructure construction, while the various parts of the private sector serve as the contractor and operational manager of commerce, profiting from these large scale infrastructure investments.

There are valuable lessons that the U.S. HSR community can learn from these European experiences Among those lessons is the need to forge new models of cooperation between banks, project sponsors, Federal, state and local governments and international sources of finance in new forms of partnerships. Such partnerships will maximize needed capital investment, bring each segment of the partnership into "ownership status" within the project, emphasize private sector operational efficiencies. Thus also these partnerships will minimize unnecessary wasted resourcer expenditures on compliance coupled with unneeded or multiple overlapping and duplicative governmental demands.

5.4. THE MULTINATIONAL EUROPEAN EXPERIENCE OF FINANCING HIGH-SPEED RAIL IN EUROPE

HSR achieved phenomenal success across much of Western Europe over the past decade and onehalf. This proven record is further propelling Europe towards even bolder plans for binding the continent together in the future. This HSR ground transport web will integrate the continent's diverse economies.

This continent, with a projected 1995 population of 385 million, *constitutes the single largest integrated market in the developed world.* HSR will serve as a vital backbone for accelerating this burgeoning mega-economy into a principal global competitor in the increasingly competitive global market. It will help define the European economy as *one integrated whole.*

The central entity for planning deployment of a "Single and integrated HSR system across a united Europe" is the Community of European Railways (COER). COER prepared and distributed the most widely respected summary of the European plans for the financing and deploying of HSR across the European continent.[7] In that plan they assert:

> As the first high-speed projects implemented (Shinkansen in Japan and South-East TGV in France) show, high speed rail services are especially appropriate in the medium-distance (130 to 700 mile distance) bracket for links between heavily-populated urban centers with highly mobile populations. Europe fits this profile *well and will do so increasingly with the prospects of the single market.* (emphasis added)

The Costs of Expanding HSR Across Europe

Across the continent "national" and continental HSR plans are under development. Recient, European HSR System experienced an expansion from the 1,700 linear miles in place in 1992 to over 2,500 by the end of 1995 alone. The continent expended in excess of $60 billion to unite Europe with high speed rail by the end of this year (1995) and in excess of $40 to $65 billion additional is projected to be spent between 1995 and 2005.

Prospects are growing for a very ambitious integration of the entire continent of Europe with a ribbon of 5,600 miles of very high speed rail linkages with speed in excess of 155 to 200 miles per hour. The plan also will include 9,300 miles of upgraded track capable of speeds of 124 miles per hour travel.[8] Costs for this transportation initiative for all EC countries could exceed $120 billion by the year 2010 for all the nations of the EC. The financial and socioeconomic profitability of these investments is explored in earlier publications prepared by this and other authors in the past.[9]

France

The French (and now other regions of Western Europe) have experienced phenomenally successful operation of the TGV in revenue service for almost 14 years of operation since the first section between Paris and Lyon opened in September, 1981.[10] The success of the southeast line is well established technically, commercially, economically and financially. The first TGV line profits will pay off its investment in less than ten years.[11]

Germany

In Germany the Federal Infrastructure Plan[12] makes provision for almost 3,000 linear miles of high-speed line. The first main north-south link consists of the Hanover-Wurzburg new line (206 miles), seventy miles of which came into service in May, of 1988. By the end of 1995, the Germans planned to have all internal passenger movements made by ground transport. This plan was designed to relieve the insatiable demands of an increasingly mobile population for air and auto travel. That goal was delayed indefinitely with the reunification of East and West Germany.[13]

Sweden

Another important model is the successful deployment and operation of the X 2000 across Sweden and the unique HSR public- quasi-private ownership and operational relationship that exists in that country. A number of other European nations are beginning to emulate the Swedish national rail realignment plan and socio-economic assessment techniques. This successful experience holds great promise for guiding HSR/Maglev development in the United States.

Spain and the remainder of western Europe

Spain proposes construction of 900 miles of new high speed rail track and prospects of narrowing the gauge of their entire existing network by 9.5 inches to conform to the European standard gauge. RENFE, the Spanish national railroad company, completed the $2.8 billion, 186 mile TGV high speed rail system between Madrid and Seville during 1992, in time for the World's Fair in Seville. RENFE is also studying the Madrid Barcelona line, a northwestern route from Madrid over the Sierra, the Guadarrma mountains to Baladolid and to connect to France and Europe generally as well. Italy and the remaining Scandinavian nations also have ambitious financial HSR plans.

How Will These High Speed Rail Developments be Financed in Europe?

One section of the COER plan reports on financing this explosion in HSR expansion:[14]

> Very different solutions may be adopted depending on the country concerned...at one extreme there is the principle whereby infrastructure development is the responsibility of the State, in particular because it is an integral and fundamental part of land use development policy: this applies in Germany, Italy and Spain. At the other extreme (if rare) view is an entrepreneurial based predominantly private management principal, such has been the case with the EUROTUNNEL.

Somewhere between these two extremes, the SNCF in France and the British Rail in the United Kingdom are responsible for their own development programs and for funding arrangements using their own resources or by borrowing money. These system still require partial subsidies, such as the EEC grant for the BR East Coast line and the 30% State participation in the SNCF's Atlantic TGV infrastructure.[15]

> Since the major advantages of building these systems accrue to the public sector it is quite reasonable for the public sector to take charge of providing large subsidies for part of the infrastructure funding (as it does for other types of infrastructure most notably *roadway infrastructure*)...so the private operators bear the brunt of the risk within limits stipulated by Government while having the support of financial aid from the State budget or of guaranteed public loans.
>
> ...it is true to say that funding entirely from private sources leaving private operators the liberty to fix their rates at the level they wish are a costly answer to the problem. Such a solution was *only possible in the case of the Channel Tunnel project* because the expected return on investment was so exceptional. (emphasis added).

As reported above, the issue of funding is far from resolved in all of these cases. The central question that remains is where will the money come from for these projects. This is apparently still an open question. One author indicates,

> "All European companies are very indebted, and they are all about to make gigantic investments. The question of whether they can continue to augment their traditional borrowing is not a question of the capacity of the markets, which are always prepared to lend money as much as the railways want, since they are backed by

governments. It is a question of whether the European countries are prepared to increase their public and private debt."[16]

European governments will keep major portions of their national railway systems in the public sector, but increasingly are moving toward quasi-private ownership and operation of rolling stock and operations. Some publicly-owned rail systems however, are opposed to private equity partners. The Director of Land Transportation of the French Ministry[17] said, "SNCF would resent it bitterly, and there would probably be a general strike if a (private) company were involved... That is a political or social question." Generally the EC is opposed to giving any kind of public guarantees to private financing, however major public subsidies are provided in various forms to all systems.

The one exception to this general rule is the Eurotunnel which serves as a prime example of potential private sector limited financing for high speed rail systems of this sort. Generally, complete private sector financing will be considered only in limited settings where the projects can be isolated such as those of tunnels. General prospects for infrastructure financing of high speed rails systems across Europe will continue to fall predominantly within the public sector.

In Chapter 6 of this book *Comparison of European and American Public Transportation Financing Policies* provides an overview of major public policy differences between Europe and U.S. on the financing and use of national transportation systems. These important differences conclude in very different mode mixes and ridership patterns as a conscious result of *explicit public policy direction*. This broader transportation policy discussion should help frame the national context within which the HSR policy decisions are formed.

Finally, local and private sector cooperation and financing will be explored in greater detail in Chapter 7, *Implementing High Speed Rail in Europe with Local Government and Private Sector Cooperation-A Case Study – Lessons from Lille* focuses on the growing importance of local government and the private sector in successful planning, financing and implementation of HSR in Europe. This section provides an important set of conclusions that are equally important for successful deployment of HSR anywhere in the United States as well.

Chapter 8 Analyses the Impacts of Florida High Speed Rail provides a profile of the private public partnership in place in Florida and the economic, environmentel transportation, saftey and energy impacts predicted from implementing HSR across central and south east Florida.

FINDINGS AND CONCLUSION

- Historically public transit and HSR are publicly funded in Europe and generally associated with provision of an essential public function—providing efficient and effective public transportation services.
- There is a growing need for a newly defined cooperative private-public relationship to finance needed European infrastructure—including proposed massive HSR infrastructure investments.
- Financing new transportation and HSR infrastructure in the future will increasingly require private sector participation as equity investors.
- The European view of financing HSR infrastructure across the EC recognizes that ridership fees and other user charges on HSR and related infrastructure will not be sufficient to finance the full cost of development of these systems.
- Increasingly, private sector principals of management and operation are being incorporated into the structure and operation of Western European national rail systems examined in this study.
- The convention is to separate the responsibilities and functions of national rail systems into public and quasi- private areas of specialization, based upon the areas each sector is most capable of performing.
- Generally the public sector retains ownership and responsibility of maintenance of the rail infrastructure. This tends to be the most expensive and resource intensive end of the system, and the segment that will not be expected to make a "profit" or cover it's costs.
- The private or quasi-private sector retains ownership and operational responsibilities of rolling stock, system scheduling and passenger and in freight operation across the system.

Across most of Western Europe national governments pledged to absorb and repay most of the national rail systems' historic debts. This allows the "new" restructured public-private ventures to start financially with a clean slate and improves their prospects for securing lower interest loans for needed upgrades and infrastructure improvements.

NOTES

1 Segments and quotes within this section were abstracted from "Financing Transport Infrastructure," *Transport in Europe-Creating the Infrastructure for the Future*, Mr. Pitt Treumann, Director General for Lending European Investment Bank, 2 & 3 March, 1993, Financial Times Converences, 6.1.
2 *The Economics and Financing of High Speed and Financing of High Speed Rail and Maglev Systems in Europe, An Assessment of Financing Methods and Results With The Growing Importance of Public Private Partnerships and Implications for the U.S*, Chapter 6, "Results of the Financial Analysis of the High Speed Train Network Across Western and Eastern Europe", Dr. Tim Lynch, Florida State University, March 15, 1995, Center for Economic Forecasting and Analysis.
3 *Ibid*, Lynch, 1995, Chapter 7 *The Critical Role of Local Government-Private Sector Cooperation in Financing Planning and Implementing High Speed Rail in Europe*.
4 *Ibid*, Treumann, 1993.
5 *Overview of the Financial Proposals for the Florida High Speed Rail and Maglev Transportation Systems, Planning Process for High Speed Ground Transportation: Lessons Learned in Florida and California*, Lynch, Thomas A., Ph.D., Prepared for the Institute of Transportation Studies and the California Department of Transportation, Assistant Director, Florida High Speed Rail Commission, Sacramento, California, January 31, 1991.
6 *Financing High Speed Rail and Maglev Systems in Europe Japan and the United States: Implications for Systems Financing in Florida*, Lynch, Thomas A., Ph.D., Center for Economic Forecasting and Analysis, Florida State University for the Florida Department of Transportation, 1992.
7 *Proposals For a European High-Speed Network*, Community of European Railways, British Railways Board (BR), Chemins de fer federaux suisses (CFFS), Societe National des Chemins de Fer Luxembourgeois (CFL), Organisme des Chemins de fer helleniques (CH), Coras lompair Eireann (CIE), Campinhos de Ferro Portugueses, (CP), Deutsche Bundesbahn (DB), Danske Stasbaner (DSB), Ente Ferrovie dello Stato (FS), N.V. Nederlandse Spoorwegen (NS), Osterreichische Bundesbahnen (OBB), Red Nacional de los Ferrocarriles Espanoles (RENFE), Societe Nationale de Chemins de Fer Belges (SNCB), Societe Nationale des Chemins de Fer Francais (SNCF), 1989.
8 *Ibid. ENR*, p. 22.
9 *Ibid*, Lynch, 1995, Chapter 2 *Results of the Financial Analysis of the High Speed Train Network Across Western and Eastern Europe*.
10 *Address of John A. Harrison to the TRB Executive Committee*, January 9, 1990, Transportation Research Board Executive Committee Meeting, Washington, D.C, (Unpublished).
11 *Ibid*, Lynch, 1995, Chapter 3 *The Economic and Financial Success of the TGV System in France*.

12 (BVWP '85–94).
13 *Ibid*, Lynch, 1995, Chapter 4 *Success of the Intercity Express and Maglev Deployment in Germany*.
14 "Financing and Developing the European Network-Different Philosophies in the Different Countries", COER, 1989.
15 Confirmed with the French Ministry of Finance, private communications with Dr. Thomas A. Lynch, Paris, France, September, 1989.
16 *Ibid*. Sychraba.
17 *Ibid. ENR*, quote of Claude Gressier.

6 COMPARISON OF EUROPEAN AND AMERICAN PUBLIC TRANSPORTATION FINANCING POLICIES

Dr. Tim Lynch,
Director
Center for Economic Forecasting
and Analysis
Institute for Science and Public Affairs
Florida State University
Tallahassee, Florida, 32306

INTRODUCTION

Historically, the financing of public transportation in Europe and the United States vary dramatically. One author[1] states,

> When visiting cities in other countries (Europe) one is often struck by differences in their transportation systems. These differences are among the most visible indicators of *variation in underlying social, political and economic systems*. There is an unmistakable increase in the relative importance of the automobile and corresponding decreases in importance of public transport modes such as bus, streetcar, subway, and commuter rail (and high speed rail between Europe and the United States).
>
> Whatever their limitations, the availability of data strongly suggest that public policies significantly effect travel behavior and explain more of the variation among countries in urban transportation than can be explained by differences in income levels."

Tabl. 6.1 provides a comparison of auto ownership levels in Eastern and Western Europe and North America, per thousand population. Note that while western Europe gained considerably on the U.S. in the past 15 years, the United States started well ahead of Europe after World War II and continues to dominate vehicle ownership per thousand population at 555 in 1987. West Germany is next with 470 per thousand population and only slightly trails the U.S. in 1994 estimates. Estimates for 1994 see this gap declining for most of the nations of western Europe.

Net ownership increases demonstrate that eastern Europe and the former Soviet Union accelerated ownership since deregulation of automobile constraints and increases in production in those countries. Tables 6.1. and 6.2. also demonstrate a leveling off in the growth in auto ownership everywhere except the former Soviet bloc countries.

TABLE 6.1. Comparison Of Auto Ownership Levels In Eastern And Western Europe And North America.

	Cars Per Thousand Population					
	1950	*1960*	*1970*	*1980*	*1987*	*1994**
Country						
Soviet Union	3	3	7	31	45	59
Poland	1	4	12	66	140	214
Hungary	1	2	10	152	157	162
Czechoslovakia	8	14	46	83	174	265
East Germany	N/A	9	61	149	207	265
Denmark	28	87	205	278	313	348
Netherlands	15	49	171	307	348	389
Belgium	32	88	200	319	349	379
United Kingdom	49	109	213	276	355	434
Austria	6	61	159	300	355	410
France	39	130	233	355	393	431
Italy	10	39	170	310	397	484
Sweden	42	160	274	360	402	444
Switzerland	32	97	214	375	421	467
Canada	140	224	306	427	454	481
West Germany	13	82	208	375	470	565
United States	268	345	428	537	555	573

Source: John Pucher, "Capitalism, Socialism and Urban Transportation, Policies and Travel Behavior in the East and West", *APA Journal*, Summer 1990. Motor Vehicle Manufacturers Association 1952 1962, 1972, 1982, and 1989, 1994 values are extrapolation of 80–87 trends.

TABLE 6.2. Percentage Change in Automobile Ownership in Europe and North America.

Country	*1950 - 60*	*1960 - 70*	*1970 - 80*	*1980 - 87*	*1987 - 94**
Soviet Union	0.0%	13.3%	34.3%	6.5%	3.1%
Poland	30.0%	20.0%	45.0%	16.0%	5.3%
Hungary	10.0%	40.0%	142.0%	0.5%	0.3%
Czechoslovakia	7.5%	22.9%	8.0%	15.7%	5.2%
East Germany	-	57.8%	14.4%	5.6%	2.8%
Denmark	21.1%	13.6%	3.6%	1.8%	1.1%
Netherlands	22.7%	24.9%	8.0%	1.9%	1.2%
Belgium	17.5%	12.7%	6.0%	1.3%	0.9%
United Kingdom	12.2%	9.5%	3.0%	4.1%	2.2%
Austria	91.7%	16.1%	8.9%	2.6%	1.5%
France	23.3%	7.9%	5.2%	1.5%	1.0%
Italy	29.0%	33.6%	8.2%	4.0%	2.2%
Sweden	28.1%	7.1%	3.1%	1.7%	1.0%
Switzerland	20.3%	12.1%	7.5%	1.8%	1.1%
Canada	6.0%	3.7%	4.0%	0.9%	0.6%
West Germany	53.1%	15.4%	8.0%	3.6%	2.0%
United States	2.9%	2.4%	2.5%	0.5%	0.3%

* Extrapolations basc extrapolation 80–87 source J. puches *et al*, 1990

Even though auto ownership grew appreciably in Europe over the past two decades, the level of dependence on the automobile is typically slightly under half that of the U.S. Public transport ridership levels are three to nine times higher in Europe. Comparatively, pedestrian and bicycle modes are also three to four times higher in Europe than in the United States or Canada. To what do we attribute these trends? Largely the conscious public policies formulated by each nation.

Explanations For Differences In Modal Share Distribution

Perhaps the most visible explanation of the relative size and dominance of the public/private sector influence on public policy are the differences in fuel costs and subsidies across these nations. The public sector as a percentage of gross domestic product in the United States is only 37%. By comparison across western Europe with the exception of Switzerland, much larger public sectors constitute between 46% to 63% of gross domestic product expenditures. These differences in relative size influence public transportation policy debates.[2]

For example the respective cost of fuel for most of the developed nations of the world vary by only 25%. Meanwhile the respective cost of fuel taxes for these nations varies by 800% or more. So while the various prices to imported oil based fuel do not vary substantially the cost for fuel but the final price is largely influenced by the addition of taxes. This is a direct reflection of public policy among nations. Per capita gasoline consumption in a sample of U.S., Canadian and European cities, provides comparison regarding the excess consumption predicated on the lower cost of fossil fuels in the U.S.

The U.S. consumption of transportation fuel is 200% to 250% in excess of its major European and Asian competitors. Much of this excess consumption is a function of inexpensive fuel. But as one author demonstrates automobile based transportation systems require high levels of private ownership while public transport systems in both socialist and capitalist countries are almost always publicly owned. Public transportation policy therefore, evolved within the foundation of the social and economic fabric of culture within each of these nations, the urban travel and subseguent behavior mirror many aspects of the society as a whole.

Regarding the relationships between culture, public policy, and transportation, John Pucher writes:

> "These differences in urban transportation...to a significant extent result from decades of deliberate public policy...Policies in the United States have strongly encouraged auto ownership and use. For many decades, large subsidies to highway construction, automobile use and low density suburban housing have made the automobile very appealing if not irresistible while...a decline of public transport... was eliminated for most Americans anyway. In western Europe... public taxation policies make it much more expensive to own and operate automobiles...Moreover large subsidies to public transport in western Europe have helped finance much more extensive and much higher quality public transport services than in the United States.
>
> The automobile, enabling almost unlimited freedom of movement and location, embodies the principals of individualism, privatism, consumerism and high mobility. By contrast, public transport fosters communalism, depends upon, planned transportation and land use systems, restricted mobility and less individual freedom of choice both in travel and in location. In the United States the automobile has become a virtual necessity of life.

> Those without automobiles in the United States are forced to suffer a substandard level of accessibility and some Americans view people without automobiles as outright deviants... In contrast to the United States, people without automobiles do not suffer from immobility (in Europe) to nearly the same degree as do Americans without automobiles thanks to generally excellent public transportation systems."[4]

Pucher asserts that in most western European countries *public transport is viewed as an essential public service* not out of ideology but rather on practical considerations such as congestion reduction, pollution abatement, compact urban development, traffic safety, energy conservation and, probably most important, the enhancement of mobility and travel options for *all* segments of the population.[5] In the United States, the vast majority of the population view public transport as totally irrelevant. At its worst it is nonexistent. At best it is the mode of last resort used predominantly by those so poor that they cannot afford to use an automobile.[6]

In the United States an individual is able to take three to four trips by automobile for the cost of one trip by public transit. By sharp comparison the cost of using an automobile in Europe is generally equal to or greatly exceeds the cost of using a public transit.

In Italy, for example, the average cost of using an automobile is 6.5 times higher than that of public transportation while in Austria the average price is 3.17 times higher, falling to a low of 0.8 of the cost of an automobile in Sweden. By comparison, the much lower typical public transit fare is half to slightly in excess of the cost of an automobile in the remainder of Europe.

By contrast in eastern Europe, the much higher publicly subsidized transit systems of eastern Europe exceed fuel costs by factors of 8.2 to 11.7 respectively in Poland and Hungary. This reflects the much higher direct subsidy to public, transit because of historically publicly dictated constraints on much higher fuel costs.

Tab. 6.3 shows that auto use in the U.S. historically exceeded that of western Europe by a factor of two or almost three. U.S. use of public transport in the late 1970s and early 1980s constituted only 3.4% of all urban passenger trips compared to 11% to 26% across Europe and Canada during (and since) that period.

TABLE 6.3. Modal Split In Urban Passenger Transport (As Percent of Total Trips)

	Mode					
		Public			*Motorcycle*	
United States (1978)	82.3	3.4	0.7	10.7	0.5	2.4
Canada (1980)	74.0	15.0		11.0		
West Germany (1978)	47.6	11.4	9.6	30.3	0.9	1.1
Switzerland (1980)	38.2	19.8	9.8	29.0	1.3	1.9
France (1978)	47.0	11.0	5.0	30.0	6.0	1.0
Sweden (1978)	36.0	11.0	10.0	39.0	2.0	2.0
Netherlands (1984)	45.2	4.8	29.4	18.4	1.3	1.0
Italy (1981)	30.6	26.0		43.4		
Austria (1983)	38.5	12.8	8.5	31.2	3.7	5.3
Great Britain (1978)	45.0	19.0	4.0	29.0	2.0	1.0
Denmark (1981)	42.0	14.0	20.0	21.0	—	3.0

Sources: J. Pucher et al., "Socioeconomic Characteristics of Transit Riders," Traffic Quarterly 35, no. 3., pp. 461-483; German Ministry of Transport, "Modal-Split," Forschung Stadtverkehr, Heft A-1, Bonn, West Germany, 1984, p. 44; F.V. Webster et al., "Changing Patterns of Urban Travel," Transport Reviews 6, no. 1, January-March 1986, pp. 49-86; Haskoning Koninklijk Ingenieurs- en Architectenbureau, Changes in Traffic and Transportation Patterns in Denmark, Netherlands Ministry of Transport, The Hague, Netherlands, 1984, p. 60; Austrian Ministry of Transport, "Verkehrserhebung–Benutzung von Verkehrsmitteln," Statistische Nachrichten 40, no. 11, p. 818; Italian Ministry of Transport, "Il Trasporto Publico Locale: Analisi per Regione," Rome, 1985, p. 11, Netherlands Central Statistics Bureau, National Travel Survey, Maastricht, Netherlands, 1986; Swedish Ministry of Transport, "Trip making by Mode in Sweden," Stockholm, 1978; Statistics Canada, Characteristics of Urban Travel," Ottawa, Canada, 1982.

A clear example of the difference in policy between the United States and Europe is seen in where West Germany total subsidies to urban public transit from all levels of government in 1988 was 20.7 billion marks or equivalent to 12.3 billion U.S. dollars. This is only slightly less than the 12.6 billion spent by all government levels in the United States to subsidize urban public transit for a population over four times larger.[7] The per capita subsidy to public transport in West Germany is four times larger than that of the United States, $202 vs. $51 in 1988 dollars.

Decline in the Cost of Gasoline in the U.S.

Another example of differences in concious national policy is the very low cost of gasoline in the United States. Figure 6.1 provides a profile of the real 1995 and nominal (price paid at the pump in each year excluding general inflation) dollar cost of a gallon of gasoline in the U.S. The in 1970s when most people remember gasoline costing $.35 a gallon its real (1995 equivalent) cost was almost $1.50.

During the OPEC oil crisis of the 1970s and early 1980s the nominal price of gasoline rose above $1.38. In 1995 real dollar equivalents the price is actually $2.45 per gallon. Thereafter prices fell back close to the nominal $1.00 level to conclude in 1995 close to $1.18 per gallon on average. In the early 1970's almost 33% of the price of gasoline was federal, state or local taxes. Unlike Europe and developed nations in Asia, inflation eroded away the value of those tax revenues. *In the United States the real value of federal, state and local taxes has fallen dramatically by over 50% of it's 1970 level over the past two plus decades.*

Trends in Public Transport Service and Ridership

While the data in Tab. 6.3 are somewhat dated, they provide an insight into historic 1960 to mid 1980s (and beyond) trends in urban

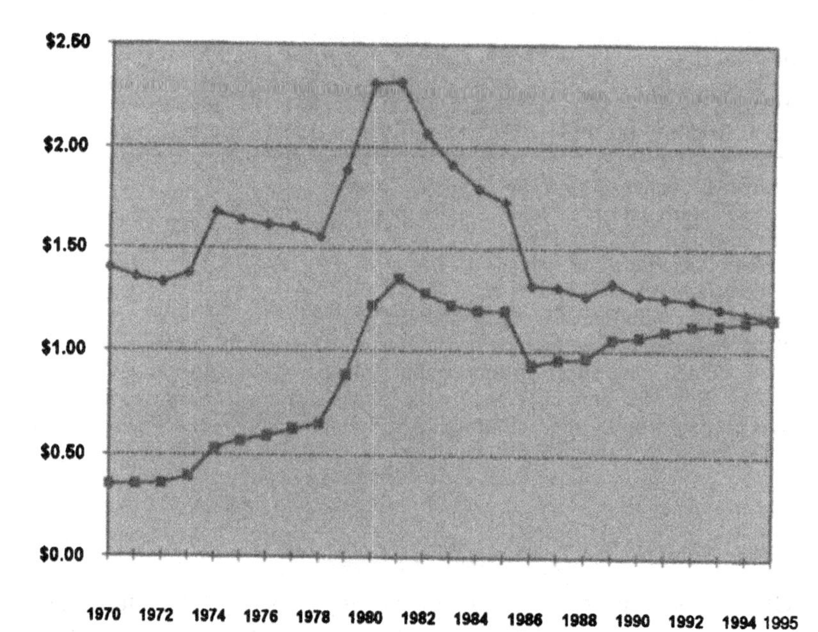

Figure 6.1. Real 1995 and Nominal Cost of a Gallon of Gasoline in the U.S. (1970-1995)

public transport use in the U.S. compared to Europe. Public transportation modal share use by urban passengers, as a percent of total trips, shows the U.S. per capita kilometer and trip use between 17% to 50% of European and Canadian levels. Average annual change in public transport use is also declining in the U.S. while generally increasing in the major European nations.

TABLE 6.4. Trends In Urban Public Transport Service And Ridership

	Vehicle-km. of Service			Passenger Trips		
Country	*Total (million) 1982*	*Per Capita 1982*	*Average Annual Change (%) 1965–82*	*Total (million) 1982*	*Per Capita 1982*	*Average Annual Change (%) 1965–82*
United States	2,132	9.4	+0.4	6,038	28	−0.6
Canada	433	18.0	+7.1	1,859	77	+3.6
West Germanya	1,922	31.7	+2.2	5,841	97	+0.8
Switzerland	104	16.3	+0.7	695	106	+0.5
Franceb	601	11.1	+2.5	2,898	84	+2.4
Swedend	475	57.1	+3.2	591	71	+2.0
Netherlandsa	115	8.8	+0.0	536	80	+1.2
Belgium	79	8.0	−0.5	372	85	−0.3
Italyc,e	624	11.0	+2.7	4,851	85	+3.7
Austriaa	124	16.4	n.a.	821	101	n.a.
Great Britain	2,113	37.8	−2.0	5,490	114	−4.1

Source: J. Pucher er al, 1990
a. Vehicle-km data for 1984.
b. Vehicle-km data for 1983.
c. Vehicle-km data for 1981.
d. Vehicle-km data for 1980.
e. Passenger Trip data for the period 1973 to 1981.
Note: In general, data exclude commuter rail. Figures for Sweden and Denmark include some short-distance public transport in rural areas.
Source: Based on data collected by the author directly from transport ministries in each country and on unpublished data provided by the Transport and Road Research Laboratory.

Public Subsidies by Transportation Mode in the United States

Of the funds that remain most of the emphasis is on the automobile. Furthermore this erosion in fuel tax value resulted in developing alternative-subsidy-funding sources for several transportation modes. Massive public subsidies to the automobile and highway construction mode have visibly slanted public transportation policy in the United States away from large scale public transit encouragement. Results from this authors research suggests[8] over the a ten year time period federal, state and local governments subsidized highway construction and operation between $3.6 to $18 billion annually in 1995 dollars.

On average, annual federal, state and local highway subsidies exceeded $14.2 billion with decade total subsidies in excess of $142 billion. By comparison, the air mode received an annual subsidy ranging from $ 1.9 to $4.1 billion annually with an average of $2.9 and a total of $29 billion over that decade while the rail modes, both freight and passenger, received annual total subsidies ranging from a low of $1.4 to a high of $8.8 billion with an average of $4.6 and a total of $46 billion over the decade (1995 dollars).

These and other massive subsidies to highway construction, automobile use and low density suburban development dictate universal automobile ownership to be the norm rather than the exception in the United States. Subsidies to auto use take many, forms some of which are so indirect that they are not even perceived as subsidies.[9] For example, the provision of employee parking as a tax free fringe benefit represents a subsidy to auto commuting estimated to range is from $12 billion to over $50 billion per year.[10]

Again, these conscious public policies fostering development of one mode and deferring development of other modes are an explicit part of conscious public policy in one nation or the other. These respective mode use rates then are not a reflection of exogenous factors such as levels of income or other socio-demographic characteristics, as much as conscious public policy.

The conclusions as pointed out by one author are that:

> These conscious public policies make automobile use in former eastern Europe conceived of impossible as and Russia almost impossible, just as transit use for most Americans if conceired as impossible for all practical purposes. Modal choice in the United States is extremely limited and massive subsidies to suburbanization and auto use over many decades in the United States, encouraged Americans to live at low densities and to rely almost exclusively on the automobile for their travel needs.
>
> Because Public transport services by contrast, are so limited in most areas they do not offer a feasible transport option. Americans chose the automobile over public transport, but only under extreme

public policies that strongly favored auto use while neglecting public transport.[11]

SUMMARY

In summary, while eastern and western Europe are generally (tilting) their transportation modes increasingly towards the automobile it is highly unlikely that they will ever reach the level of automobile dependence that the U.S. has acquired. Their emphasis on high quality public transportation will keep these modes competitive and viable even as many Europeans increasingly approach the levels of automobile mobility enjoyed by the United States. These shifts in mobility in Europe will also generate, as they have in the United States, increasing demands for automobile related consumer goods that will further stimulate the consumer oriented economies of western Europe.

A second and related, perhaps increasingly important questions for the United States are: Will a reciprocal balance and shift of modal share towards public transportation evolve as automobile and air travel saturation set in; Will higher levels of quality service for public transit become a public policy priority across federal, state and local levels of government as saturation point in automobile ownership, congestion, energy consumption and pollution become a marked threat to the quality of life enjoyed by American urban citizens? The later can only evolve when *conscious public policy explicitly direct the incentives* ***toward*** *public transportation and* ***away*** *from the high automobile subsidies the U.S. automobile modes has enjoyed over the past five decades.*

FINDINGS AND CONCLUSIONS

- Public sector explicit and implicit policies decisively dominate the decisions about which modes will dominate that nations transportation infrastructure.
- Public sector policies will direct the decisions regarding the amount of subsidies each transportation mode will receive
- All transportation mode, including high speed rail systems, in each countries examined are publicly subsidized.
- The size as well as persistence of these subsidies vary widely dominated by internal and external cultural and environmental factors, across modes and nations.
- Public subsidies for support of the automobile exist in each country examined (except historically across former Eastern Europe) but are highest in the U.S..

- In virtually all of Europe all income levels use public transportation services.
- Public subsidies in Europe are considered good investments in the provision of important public services – high quality person and goods movements and are broadly supported by all income groups.
- With the exception of several major cities, public transportation use in the U.S. is dominated by low to moderate income travelers.
- Subsidies for public transportation in the U.S. are often considered welfare support for public "transportation dependent" lower income segments of urban society.
- Public transportation and high speed rail subsidies are most favored across former Eastern and Western Europe. Public subsidies for mass transit in the U.S. are considerable but public subsidies for passenger rail are very modest compared to European standards.
- The United States subsidizes the automobile well beyond any other nation in the developed world.
- Over the past five decades the United States powerful private sector industrial interests in the United States reinforced cultural bias to influence public policies and subsidies to advance the automobile mode and to de-emphasize support for public transit and (conventional and high speed) rail modes.
- These dominant interests combined to under value public transportation and under price automobile ownership and operation costs. (For example the real price of gasoline in the U.S. is less expensive today than any time since the end of World War II.)
- Decades of under investment in public transit and rail in the U.S., massive subsidies for the automobile and the dominance of pre-committed land use decisions have developed into an almost intractable situation. In many cases this undermines the viability of deploying high volume public transportation fixed rail systems.

NOTES

1 John Pucher, "Capitalism, Socialism and Urban Transportation, Policies and Travel Behavior in the East and West," *APA Journal*, Summer 1990.
2 *Ibid.* Pucher, 1990.
3 *Ibid.* Pucher, 1990.

4 *Ibid.* Pucher, 1990.
5 European Council, Ministries of Transport, 1984, 1986.
6 Bruce Briggs, 1975. Mass Transportation and Minority Transportation the Public Interest Summer 42–74.
7 West German Ministry of Transport 1989a, Bericht..Jahr 1988, Bonn, West Germany Ministry of Transport (**TL: FILL IN FOOTNOTE**) as referenced in John Pucher.
8 Lynch, T.A. "Public Transportation Financing and Subsidies by Mode in the United States," prepared for High Speed Rail Association Conference, Anaheim, CA, May 7, 1991. Florida High Speed Rail Transportation Commission.
9 *Ibid.* Pucher, 1988.
10 Meyer, J. and Gomez Ibanez. 1981 "Autos Transit and Cities," Cambridge, MA, Harvard University Press, Stuber, M., D. Shoup, and M. Wachs. 1984 "Affects of ending driver paid parking for sole drivers." Transportation research record 957:46–54 as referenced in Pucher, 1990.
11 *Ibid.* Pucher, 1988.

7 IMPLEMENTING HIGH SPEED RAIL IN EUROPE WITH LOCAL GOVERNMENT AND THE PRIVATE SECTOR COOPERATION

Dr. Tim Lynch,
Director
Center for Economic Forecasting and Analysis
Institute for Science and public affairs
Florida State University, Tallahassee, Florida, 32306

A CASE STUDY—LESSONS FROM LILLE

7.1. INTRODUCTION

This chapter first describes the nature and possible public-private partnerships in joint HSR ventures. Next, the background to applying a specific partnership in Lillp is developed. Financing the chapter concludes with the important local government-private sector relationship that developed between the high speed rail (TGV) developer and the Ville de Lille in France. This partnership concluded in successful deployment of the TGV system into the heart of Lille and serves as a model for local governments anywhere in the world hoping to attract and retain a high speed ground transport system to service their community. A number of important conclusions that may help guide HSR development in the United States can be drawn from this Lille Case Study. Among the most important is recognizing and agreeing to critical public-private (or quasi-private) partnership responsibilities in planning, financing and implementing HSR systems. This division of labor can vary some what from site to site, but the essential components are fairly well agreed to in practice.

7.2. PUBLIC / PRIVATE RESPONSIBILITIES IN A PARTNERSHIPS

The Public Sector

The public sector is primaily responsibility for funding, owning and operating the infrastructure and providing the means to complete needed system design and mitigation planning and development. These tasks are distributed among federal, state and local governments with increasing funding responsibilities at the federal level and land use planning and controls and environmental permitting and mitigation at the state, regional and local levels.

The Private Sector

The private sector or quasi-public-private sector can be classified as the entity that owns and operates the HSR system in a "business like" manner with investment requirements, such as a threshold investment criteria (like the SNCF 8% criteria), before a HSR investment is committed. This public-private entity begins to make "traditionally government" institutions into much more efficient "market responsive" institutions and reduces the traditional governmental excess and waste that sometimes accompanies public rail development.

The fully commercial private sector (in cooperation with the public and quasi-private sector) responsibility to; efficiently secure and integrate the terms and conditions of financing (as a facilitator but not a fully responsible debtor for repayment of debt); provide for initial land use planning and design and; in a number of cases, become the operator of the rail system in a profit driven framework. This model assumes development of sufficient public subsidies (such as in France, Sweden and elsewhere) to ensure financial stability of viable routes over the initial period of greatest need.

Differences in US and European Partnership

A unifying and integrating sense of cooperation pervades the public/private ventures across Europe in developing HSR projects compared with those equivalent equivalent unsuccessful attempts to date within the U.S. In the U.S., the members of the private and public teams ask, "Who is going to do something for me? Who will pay my (and others) financial transportation bills? Who will provide me with revenues, regionally and locally demanded person and goods transportation services? Who will provide me with technological advances, environmental benefits, environmental damage mitigation, financial risk mitigation and a number of additional locally based needs?"

In Europe, the combined public/private HSR development teams comprehensively see what needs doing. They efficiently sit at the table as full partners, divide up responsibilities on the basis of need and specialization brought to the partnership, and get the job done. Differing levels of government do not pose purposeful restraints, but render meaningful solutions to pending constraints. There is a general recognition of the significant regional and national values in providing the levels of public transportation services, economic growth, reductions in pollution generation and energy consumption. There is also an agreed upon acceptance and high value placed on the considerable private sector profitable gains from final deployment of a successful HSR services to a region.

This becomes the general unifying single purpose for the integrated HSR public-private partnership. There is a recognition that there are sufficient (and significant) gains to all members of the this public-private alliance to warrant completion of the HSR system. The implicit recognition is to complement and not drain the system's energies and resources for parochial or individual team member needs. In the U.S. these include extracting system resources to resolve local transportation service demands, solve local environmental conflict problems, satisfy local tax revenue shortfalls or serve as the primary economic recovery and job creation engine for a city or region.

A number of these potential benefits may evolve in part or in whole from deployment of these systems. However burdening the system with resource draining expectations or extractions to achieve such ends as a condition precedent for permission for support for deployment of such a system sounds the death knell to prospects of the system's ultimate success.

7.3. THE EVOLUTION OF LOCAL AUTHORITY AND RESPONSIBILITY IN FRANCE—BACKGROUND TO LILLE HSR PROJECT DEVELOPMENT

Throughout the 1960s, 70s and 80s, France and most of the rest of Europe proceeded with strong central planning and centralized decision-making. Some scholars suggest that the decade of the 80s and early 90s experienced a substantial shift towards decentralization across Europe with the focus of urban policy decision making and planning forming at the local level.[1]

The Economic context of the Times

Increasingly, national and global economic competition work together to significantly affect local governments incentive optious and steal away those powers traditionally leveraged by the federal level in these western European governments. Two factors conspire to intensify the competitive pressure between urban areas across Europe and increasingly dominate the western European public policy decision making.

The first factor is the enhancement of competitive pressures brought by new technologies such as the development of the high speed rail technologies, new computer technologies and advances in communications. The second related, but more dominant factor, is the increasing globalization of production and markets. Increasingly, goods and services production and distribution can take place virtually anywhere. These factors, combined with the increased domin ance of the service sector in our modern economies and the multinational nature of contemporary economic activities, render[2]old notions of urban settlement and economic development moot.

As Levine points out, with the development of the European community, suppression of customs and tariffs and the construction of the English Channel tunnel, it became almost as easy for a firm to site its headquarters and production facilities in London, Frankfort, Amsterdam or any other EC city as in Paris or the French provinces.[2] This reality, combined with a series of coterminous events shook the foundation of European economic development in the 1980s and into the 90s, and resulted in a shift in public policy and shifts of power from federal to local governments. A similar phenomena evolved simultaneously in the United States under the Reagan-Bush administration with the "new federalism" of the 1980s and early 1990s. While these shifts were similar in some ways, French decentralization came with associated **increases** in federal funding while the shift of increased responsibilities to local governments in the U.S. typically concluded with *fewer* resources.[3]

The Political Temper of the Times in Lille

These were times of ideological tempering of political extremes across Europe as Communism crumbled in the East and the Soviet empire was in disarray. For example, in 1991, the Socialist Party under Francois Mitterrand came to power as a party of government in France and not the party of opposition. During that period Socialists who had historically ranked with their Communist allies in attempting to insist on government-dominated means of production across France and much of Europe began to moderate these ideologies. The Soviets were pursuing a course of economic modernization and embraced new means of cooperation and partnership between government and the business community. So also, many local European Socialists began enthusiastically embracing the right-wing free marketeers and sought to secure joint grant assistance to firms to assure creation and preservation of local jobs.[4]

While French municipalities are far less subject than their U.S. counterparts to fiscal cutback pressures, they are still less reliant on their own sources of revenue than are U.S. cities. French urban areas do not benefit as clearly as American urban areas from commercial economic development. For example, French cities do not collect new local property based tax revenue from increases in building values. They do, however, capture increases in economic expansion through job and income expansion and still are significantly benefited by economic growth.

They are therefore increasingly aggressive in pursuing economic expansion and development of future growth. This momentum combined with a renewed emphasis of capturing federal funds motivated the local public sector in Lille. This combined with the movement of international private sector interests that recognized the benefit of joint venture activities of this scale resulted in development of a Lille public-private joint ventures where both parties recognize mutual benefit and gain from the association.

Public assistance for traditional service programs, such as extensive social welfare and related services, are borne by the federal government in France and are not the burden of local government. As a result, the French system of decentralization of the early 80s did not adversely effect local governments nearly as significantly as those simultaneously transpiring in the United States and England. Under decentralization during this period, central government transfers to local governments in France actually **increased** compared with the **marked reduction in revenue** in the United States and Britain during their decentralization initiatives over the same period.[5]

During this period, local governments in France and across much of Europe gained increasing sources of autonomy and additional sources of revenue. For example, decentralization loosened restrictions on local taxation and resulted in transfers of tax levies such as property sales and automobile registration from the state to local governments.[6]

TABLE 7.1. Sources of Local Revenue in France

Year	*Local Tax Revenue*	*Federal and State Transfers*	*Loans*
1980	35%	43%	11%
1988	45%	34%	11%

Another researcher suggests that local government sources of revenue shifted significantly during the period of 1980s with renewed emphasis on local sources of revenue. This was true even though substantial sources of revenue continue to come from transfers from the federal and state levels. The preceding table indicates that local sources of tax revenue including those that were transferred from the federal level grew by 10% over the period 1980 to 1988 while federal and states sources of revenue declined by an almost equivalent amount.[7]

These increased sources of revenue also resulted in transfers of power and local decision making. These shifts of revenues and authority from the federal government also resulted in shifts to local government of new responsibilities for creation of economic growth and job opportunities. A perfect example of this shift is in the securing of financing and local administrative involvement for deployment of high speed rail in Lille (the Europeanization of Lille).[8]

7.4. CASE STUDY OF LILLE, FRANCE—BLUEPRINT FOR LOCAL GOVERNMENT IN SECURING HSR SERVICES

"Eura Lille today represents 5.3 billion francs invested. 3.7 billion francs are in the form of private investments originating from outside the region for the most part and which would neither have been brought into the Nord-Pas de Calais region without a project on the scale of Eura Lille. Public and parapublic commitments alongside Eura-Lille account (amount 1.6 billion francs) are mainly devoted to transport infrastructures. Well thought out task sharing between the public and private sectors is the strength behind the operation which is a shining example of urban projects. The Lille metropolis stands out as one of the cities which will have a key role in the Europe of tomorrow."[9]

The city of Lille represents a key component of the new wave of realities facing high speed rail financing and development across Europe. Traditionally the financing of HSR alignments across Europe came from either the Federal government or from the nationalized federally subsidized rail system. In the future financing a HSR system in western or eastern Europe or the United States will warrant significant involvement from local municipal governments and increasing partnership and financial risk sharing with the private sector.

Lille, once the stronghold of communist reactionaries, because the model of the growing emphasis towards two 1990s phenomena that will prove instrumental in the success of financing high speed rail across Europe and into Eastern Europe and beyond:

1. Dominant roles of political and financial leadership by urban municipal governments in defining and financing high speed rail systems in urban areas; and
2. Closer relationships between public and private sector investment interests in bringing these massive infrastructure high speed rail projects to fruition

Development of Private/Public Joint Venture

Lille is the fourth largest metropolitan area in France, with a population of 200,000 with a metropolitan area exceeding 1 million persons. Lille is located on the northeast border of France close to the border with Belgium and closely linked to Britain historically. Lille is also proximate to the port of Calais seventy miles away and located directly across the English Channel from Folkstone, England. Lille is what we in the United States would call a "rust-belt"city, traditionally dominated by textiles, steel works, ship building and other heavy industries.

The city's economy was hard hit during the post-industrial period of the 1960s, 70s and 80s and was estimated to be losing between eight and 10 thousand jobs a year in the textile industry alone. Much like the rest of the United States and Europe, this area was losing economic productivity to low cost wage production centers in Asia while competing with other parts of Europe and the United States for what remained of industrial development. Levine reports that the closing of mines meant the disappearance of an additional 200,000 jobs in the Lille region. Another 30,000 steel and metal working jobs were lost in the 1980s. Low wage competition from factories in liberated eastern Europe posed yet another threat to this industry.[10]

Unemployment averaged considerably more than France as a whole and the rest of the northeast region of Europe, standing at well over 13% in the early 1990s. As identified above, the key transformations of integration of Europe, increasing globalization of markets and the dropping of trade barriers were the central premises of the evolution of a new order across Europe and specifically within the urban sector of Lille. The deterioration of the economy led the municipal government and private sector to realize that new direction and new partnerships would have to be forged to move the urban area into economic recovery.

Local Urban Leadership Forges Public/Private and High Speed Rail Financing Relationships

The dominance of a few key individuals including Lille's long-time Socialist mayor, Pierre Mauroy, was pivotal in forging private/public partnerships from formerly hostile Socialist, communist and private sector interests in the city of Lille. **The municipal government brought together regional and national and private sector funding** that became the basis of financing high speed rail capital investments in the Lille region and moved the metropolitan economy forward.

Mauroy was instrumental in convincing the French rail system to locate the new TGV station in downtown Lille instead of in the neighboring countryside. His leadership also proved instrumental in securing cooperation of bankers and officials of small communities in the region. The influence Mauroy enjoyed was helped considerably by the fact that under French political rules he served in a number of joint capacities during this period including that of Mayor and simultaneously for a time, Prime Minister of the French government.

Additionally, despite criticisms from neo-Stalinists in Lille opposed to development plans, the local left-wing political arm of the Socialists and communists parties abandoned old style socialism for a new socialism that accepted the need for public/private partnerships and embraced the bank sponsored tunnel projects (to England) and additional new investments by big business.[11] This transformation of Lille's socio-political left-wing neo-Stalinists to a cooperative venture with the right-wing private sector entrepreneurial interests led to a reawakening of the need for cooperation in the urban areas. Recognition of this need also leads the way towards future economic expansion for deployment of high speed rail infrastructure investments as well as cooperation for the betterment of the municipal and urban areas collectively.

This same reawakening in the United States has yet to come to fruition. Integration of public and private sector needs are not converged to dominate the political horizon to congeal financing high speed rail development anywhere in the U.S.

The Florida FOX-FDOT partnership described in Chapter 8 is hopefully the exception to this finding. The leadership exhibited by leaders in the private and public sectors are forging a unique partnership that will serve as a valuable model for other regions of the U.S. to build on. There is a growing need for both the private and public sectors to recognize respective strengths, and operate cooperatively with a division of labor that builds on those strengths and complement respective weaknesses. The public sector must provide the institutional support of land use and environmental assessment and mitigation and financing of infrastructure. The private and quasi-private sector meanwhile must focus on financing efficient operation and management of support facilities and adjoining commercial investments. These kinds of joint ventures are essential for successful deployment of high speed rail anywhere in the United States.

Convergence of Forces to Bring HSR to Lille

Three simultaneous events conspired to bring the city of Lille into focus for this new urban growth. First, the European community suppressed customs and duties and created a genuine economic common market. Since Lille offered historic economic ties of trade and commerce with both Belgium and northern Europe and England, it's proximity to these national frontiers and the opening of trade enhanced Lille's natural locational advantage.

Second, and further accelerating this advantage was the construction of the "Chunnel" or Eurotunnel which linked England to Nord-Pas de Calais at the port of Calais, 70 miles from the center of Lille. This locational advantage placed Lille as the largest urban area closest to the point where the European continent would be joined with the British Isles.

Third, Lille served as the centerpiece and intersection of two northern European TGV lines, the TGV line from Paris to London and from Paris to Amsterdam, Brussels and Cologne. This resulted in Lille being only one hour from Paris with TGV connections, two hours or less from London by way of the Chunnel, a half-hour from Brussels and two and a half hours from Amsterdam.[12]

Combined factors of central location and intersection between urban areas, TGV location, the shifting of political power to the local level and the dominant urban leadership provided by it's Mayor poised Lille to become a boom town of Europe, fostering the recovery of the Lille's post-industrial economy.

The new TGV rail station and the area just adjacent to that station (which is close to Lille's central business and historic centers) became the focal point of this economic expansion. The station's center project is commonly referred to as EuraLille. It encompasses 120 hectares (300 acres) including 250,000 square meters (2.3 million sq ft) or more for office space, commercial service and communications center.

The municipality of Lille and the region of Nord-Pas de Calais initiated a number of large-scale projects to capitalize on the city's fortuitous location and timing. Lille initiated multiple public/private joint ventures that ensured not only the success of financing the high speed rail train station but also facilitated the success of the adjoining privately funded local commercial areas. While the investments were to bring the station to the downtown location, French TGV planners had originally indicated a desire for locating the station in the suburban areas. The local private/public partnerships viewed the station in the downtown area as central to the "economic reconversion" of the metropolis. This private/public Lille partnership were ultimately able to provide the TGV builders with sufficient financial and other concessions to successfully convince them to relocate the station to the central Lille area."[13]

Development Of A Metro Infrastructure To Assist TGV Multi-Modal Capabilities

In Lille, public-private interests combined efforts to create a high tech "technopole" to attract and implement new economic growth. In 1986, in an attempt to diversify the economic service base of Lille's high tech capabilities, three universities with combined student population of 43,000 were created in the town of Villeneuve d'Ascq. This became a prominent technology oriented center to combine education, research and industry to promote technology oriented innovation. As many as 1,800 firms including Matra, Digital, Olivette, and Hewlett Packard, with 2,300 researchers were linked to Lille's downtown urban area by a fully automated subway system developed in cooperation with the Universities of Science and Technology of Lille and the firm Matra.

Lille's metropolitan area now crosses the border into Belgium. To further facilitate this integration of international linkages, the Lille Metro plans to extend their rail linkage into Belgium proper. **This linkage, given its international character, will be co-built with joint EC financing initiatives.** This is been is designated as part of the three country Euroregion of 15 million people and it is deemed to be a significant point of future economic development. Table 7.2 and 7.3 provide a profile of the nature of the projected buildout in square meters of the Lille development surrounding the transportation station.

TABLE 7.2. Lille Station Center Project: Planned Space Allocation

	1993	Post 1993		Total Combined	
	Minimum Projection	*Minimum*	*Maximum*	*Minimum*	*Maximum*
Commercial and Service	50,000	15,000	25,000	65,000	75,000
Offices and Business Services	55,000	185,000	240,000	295,000	425,000
Hotel and Lodging	20,000	60,000	60,000	80,000	80,000
Housing	25,000	15,000	30,000	40,000	55,000
Other Facilities (Convention Center, Management etc.)	0	60,000	80,000	60,000	80,000
Higher Education	0	15,000	15,000	15,000	15,000
Total (square meters)	150,000	335,000	435,000	485,000	730,000

Source: Ville de Lille (1989)

Eura Lille Public/Private Joint Venture Financing Arrangements

The opening of the channel tunnel in mid-1994 opened the TGV rail network through Lille to serve over 70 million inhabitants in the London, on Amsterdam, Cologne, Paris quadrangle as the gateway to the entire northern European market. Lille serves as a northern European TGV juncture and will serve 30 million passengers through its two railway stations each year. The city of Lille developed a 288 acres military site in the center of Lille and expended over 5.3 billion francs, 32% (1.6 billion francs) of which are public funds and 68% (3.7 billion francs) are private funds.

TABLE 7.3. Eurail Key Financial Statistics Private/Public Joint Venture

Capital	50 million francs held by Public sector shareholders 53.95% Private sector shareholders 46.05%
Investment	5.3 billion French francs 1.6 billion public funds 3.7 billion private funds
Total Area	173 Acres in its initial phase but ultimately extending to 288 acres
Offices	45,000 square meters
Retail	31,000 square meters
Residential	Nearly 700 homes in first phase
Open Space	24 acres of urban park
Hotel Accommodations	3 hotels, luxury (3 star and 2 star), as well as self-catering and service departments for business people and students
Exhibitions/Conferences	Lille-Grand Palais: 20,000 square meters of exhibition space, meeting rooms plus 3 flexible conference amphitheaters able to hold respectively 350/500/1,500 delegates
Entertainment	2 concert halls: one with 5,000 seats in Lille-Grand Palais and others with 700
Parking	6,000 spaces in 3 locations
Construction Site	1.5 million cubic meters of earth removed, nearly 2,000 people employed at peak time
Eura Lille's team	30 Eura Lille personnel plus 450 external consultants, e.g. architects, engineering firms, surveyors and so forth

Eura Lille the Company

In 1988, the public-private French joint venture Eura Lille Metropole was established as a private company to conduct surveys into the feasibility of the expansive development of a high speed rail TGV station. Their primary focus was on developing the TGV HSR station and thereafter capturing the increased value from the commercial development in the area surrounding the station. In conformance with French law the Socit de Economie Eixte requires the majority of capital to be owned by a public sector body. The public partners own 53.9% of the company. The owners include the cities of Lille, La Madeline, Roubaix, Tourcoing, and Ville Newvu de Ascq, La Communaute, Urban de Lille (Administrative body for the Metropole area), Department de Nord (Department of the North) and Region Nord Pas-de-Calais regional council.

The private sector shareholders, including leading French banks, several international banks from Belgium, Japan and Italy, own

46.05% of the capital. Other private sector owners include regional banks and SCTEA, a subsidiary of French railway company SNCF: Lille Roubaix-Tourcoing Chamber of Commerce as well as regional insurance companies. See Table 7.4. for a complete profile of public

TABLE 7.4. Euralille: Public Private Capital Structure

Public Sector	*Percent of Ownership*	*Investment in Francs*	*Investment In Dollars*
Ville de Lille	16.4%	Fr8,190,000.	$1,365,000
Ville de la Madeline	2.5%	Fr1,250,000.	$208,333
Ville de Roubaix	2.5%	Fr1,250,000.	$208,333
Ville de Tourcoing	2.5%	Fr1,250,000.	$208,333
Ville de Villeneuve d'Ascq	2.5%	Fr1,250,000.	$208,333
Communaute Urbaine de Lille	16.5%	Fr8,235,000.	$1,372,500
Department du Nord	5.6%	Fr2,775,000.	$462,500
Region Nord Pas-de-Calais	5.6%	Fr2,775,000.	$462,500
Subtotal	54.0%	Fr26,975,000.	$4,495,833
Private Sector			
Chamber of Commerce and Industry of Lille-Roubaix-Tourcoing	3.0%	Fr1,500,000.	$250,000
SCETA (SNCF)	3.0%	Fr1,500,000.	$250,000
Caisse des Depots et Consignations	7.3%	Fr3,665,000.	$610,833
Credit Lyonnais	7.3%	Fr3,665,000.	$610,833
Dupont Scalbert Bank	7.3%	Fr3,665,000.	$610,833
Peoples Bank of the North	7.3%	Fr3,665,000.	$610,833
The Indosuez Bank	4.4%	Fr2,215,000.	369,167
Credit du Nord	1.0%	Fr500,000.	$83,333
National Bank of Paris	1.0%	Fr500,000.	$83,333
Northern Region Agricultural Mutual Bank and Credit	1.0%	Fr500,000.	$83,333
The Bank of San Paolo (Italy)	1.0%	Fr500,000.	$83,333
The Bank of Tokyo (Japan)	0.7%	Fr350,000.	$58,333
The General Bank of Belgium	0.7%	Fr350,000.	$58,333
North Regional Bank	0.7%	Fr350,000.	$58,333
Lloyd Continental	0.2%	Fr100,000.	$16,667
Sub Total	*46.1%*	*Fr23,025,000.*	*3,837,500*
Total	*100.0%*	*Fr50,000,000.*	*$8,333,333*
Initial Franc Investment level	Fr50,000,000.		
Franc to Dollar Conversion rate	6		
Initial Dollar Investment level	*$8,333,333*		

and private sector ownership and initial investments in francs and dollars. While some public and private sector initial investments were small, their involvement itself and not its magnitude was the important issue.

The TGV Rail Network And Lille Europe Station

The services and linkages offered by the TGV Lille station include Lille to Brussels 30 minutes, London 2 hours, Amsterdam 2 hours, Cologne in 2 hours and Leon in less than 3 hours. High speed trains will leave Lille Europe to bypass Paris towards the southeast, the south of France and southern Europe. There will also be direct connections to Roissy-Charles de Gaulle Airport.

Service initiated in 1994 include eight TGV trains bound for London, eight bound for Brussels, 15 bound for Paris and up to 20 direct trains bound for southern Europe each day. Since 1996 nearly 100 TGVs stop at Lille each day with as many as 30 million passengers per year passing through Lille. The Lille TGV station is expected to handle an estimated 15,000 passengers per day.

Lille is a multilevel transportation station while the Lille Flanders station will continue to serve the internal French railway network and is expected to handle 70,000 passengers a day and is located next to the Eura Lille development. Lille Europe will also include direct connection to the local underground VAL driverless metro system and tram network. The VAL (light automated vehicle system) is a computer controlled driverless metro designed and built entirely in Lille. Launched in 1983, VAL was the first system of its kind to be fully automated in the world.

Additionally, trains, tram and bus networks intersect this multimodal transportation HSR hub. The tram has been rerouted to include Lille Europe and the Port de Romarin area. Lille is also connected by direct route to A1-A25 motorway links from Dunkirk to Paris and A22 to Brussels. By air Lille is also well served by Lesquin Airport less than six miles from the city's center.

7.5. SUMMARY AND CONCLUSIONS

The conclusions to be drawn from this joint public/private Lille/Eura Lille association are that there is an essential need for local municipal public and private leadership in shaping the financing, planning and development of any HSR system within and across urban areas. The federal, regional and local public sector must be the predominant source of funding for the rail infrastructure.

The second conclusion is that a quasi public-private entity (like the SNCF in the case of Lille) must exist to establish and operate within "business like" efficient decision making criteria for deployment and operation of the HSR system. In this case if the internal rate of return must exceed the minimum SNCF HSR investment threshold of 8% to warrant provision of service. Knowing these conditions, the city of Lille (and other regional or local governments) make concessions or provide other financial awards to the HSR operator to bring the project up to the minimum acceptable rate.

The third conclusion is the need for a primary leadership role for the private sector as a primary source of planning, financing and coordination for HSR deployment within the municipality. The private sector is a critical team member within a joint venture of this sort to develop funding and plan for commercial, retail, residential, hotel and open space accommodations, exhibition and conference areas, entertainment and a variety of other commercial properties. Together these interests form a public-private corporation that (with support from the state and federal levels of government) leveraged substantial leadership for successful planning and implementation of TGV HSR across the Lille urban area.

Jointly these partners planned and developed the TGV station to serve as the focal point for development of a highly integrated multi-modal transportation system and prized private sector commercial development. The solid cooperation of all levels of government with this private-public partnership was essential to finance and deploy high speed rail in the City of Lille and will increasingly be critical to future HSR deployment everywhere else in Europe as it will in the U.S.[15]

Important Attributes of European public-private HSR ventures:

- Broad based Federal, state and local government financing is essential to implement HSR infrastructure systems in Europe and elsewhere.
- Bipartisan (or multi-party) political support across a wide spectrum of municipal governments affected by HSR is central to successful implementation of HSR systems.
- Robust and mutually beneficial public-private partnerships are essential for successful deployment of HSR systems.

- Private sector involvement is increasingly essential in financing and planning the integration of the HSR system into the commercial fabric of the urban areas.
- Each of the public-private partners performs the functions most appropriate to their unique strength to accomplish final HSR implementation success.
- Success of the HSR system ridership is increasingly tied to the system's multimodal linkages for efficient passenger transfers and mode share transfers.

National Regional and Local Government Partnership Responsibilities Include:

- Overall planning and systems design and integration with national multimodal transportation network.
- Lead responsibility for environmental planning design and mitigation strategies.
- Primary responsibility for financing, planning and implementing and maintaining rail infrastructure, including corridor alignment purchase, design and construction.
- Facilitate administrative processes that authorize and encourage successful public/private partnerships essential to implement HSR systems within and across the nation. This may include:
    - authorization for unique and favorable trading and service provision status including ownership of a sole HSR franchise award;
    - favorable government contracting authority (e.g., special indemnity, protection, transport of public cargo, etc.);
    - access to preferred lower cost public sector financing;
    - provision of credit for environmental offsets resulting from HSR deployment over the alternative modes; and
    - facilitate research and design projects that are most appropriate to accomplish the needed tasks.

Private and Quasi Private Sector Partnership Responsibilities Include:

- Establishment and methods to operate within well developed "business like" practices such as establishment of minimum investment criteria (like the SNCF 8% investment threshold) before a HSR investment is committed.

- Assisting "traditionally government" institutions to function in much more efficient quasi private "market responsive" fashion as a quasi-private entity operator of the rail system in a profit driven (including social profitability) framework.
- Assisting government in reducing governmental excess and waste that has traditionally accompanied public rail development.
- Efficiently securing and integrating the terms and conditions of financing (as a facilitator but not a fully responsible debtor for repayment of debt).
- Assisting government in providing a vision ownership and a plan for initial land use design and urban integration.

NOTES

1 "The Transformation of Urban Politics in France: the Root of Growth Politics and Urban Regimes," Myron A. Levine, Albion College. *Urban Affairs Quarterly*, Vol. 29, No. 3. March, 1994, 383–410, 1994, Sage Publications, Inc.
2 *Ibid.*, Levine, 1994.
3 "Financing High Speed Rail and Maglev Systems in Europe, Japan and the United States: Implications for systems financing in Florida." Dr. Thomas A. Lynch, Director for Center for Economic Forecasting and Analyses, Florida State University in cooperation with the Center for Urban Transportation Research (Cutter, University of South Florida), January 12, 1992.
4 Keating, M. 1991, "Local Economic Development Politics in France, *Journal of Urban Affairs* 13(4), 143–159.
5 *Ibid.*, Levine, 1994.
6 Bernier, L.L., 1990, "Policy Implications of the Decentralization Reforms in France," paper presented at the annual meeting of the Urban Affairs Association, Charlotte, NC, April, 1992, as reported in Levine, 1994.
7 *Ibid.*, Bernier (1992).
8 *Ibid.*, Levine, 1994.
9 Eura Lille, *The 21st Century in the Making*, Pavillion Souham, 44 Rue de Vuex-Faubourg, Lille Cedex, Franc. Spring, 1994.
10 *Ibid*, Levine, 1994.
11 Ardagh, J., 1989. "Lille Gets Ready for 1992," *New York Times Magazine*, December 3, 58FF. As reported in Levine, 1994.
12 *Ibid.*, Levine, 1994.
13 Ville de Lille, 1989, p. 9.
14 Source: Eura Lille and its Transportation Network, Eura Lille, 1994.
15 Lille Metropolitan Area Investment in Lille Metropolitan Area, Europe's North Star, Lille Metropolitan Area, City of Lille, 1994; Lille Metropolitan News, The Economic Newsletter of Lille Metropolitan Development Agency, Apim, June 1993; Lille Metropolitan Area, Apim, April 1994; The Atlantic TGV, The New Lines Railway Equipment, SNCF, January 1989; The TGV Atlantique, Construction of the New Line, Another Very High Speed Rail Line, the New TGV for 25 Million People, SNCF, March 1989; 55 Hectares

Au Service de la Logistique, 136 acres at the disposal of logistiques, Apim, September 1993.; National Master Plan for High Speed Rail Services, Directorate for Surface Transportation Republic of France Liberty Equality Fraternity Minister of Works Housing and Transport, Official Journal of the French Republic, 2 April 1992; The Eura Lille Newsletter, March 1994–No. 5.

8 Executive Summary: An Analysis of the Impacts of Florida High Speed Rail

8.1 INTRODUCTION

This Executive Summary presents the findings of two studies: the first entitled *An Analysis of the Economic Impacts of Florida High Speed Rail* and the second entitled *Travel Time, Safety, Energy and Air Quality Impacts of Florida High Speed Rail.* The studies were undertaken during the first half of 1997 by Tim Lynch, Ph.D., Director of the Center for Economic Forecasting and Analysis (CEFA) at Florida State University (Tallahassee) and Steven Polzin, Ph.D., Deputy Director, Center for Urban Transportation Research (CUTR) at the University of South Florida (Tampa). Two companion technical reports provide more detailed discussions of the methodologies and findings of the studies.

The Florida high speed rail project will have a significant impact on the state of Florida, creating thousands of job opportunities, stimulating economic development, serving and encouraging tourism, and providing increased transportation capacity. Travelers will save time, cost, and energy while traveling on a safer and less polluting mode.

- The diversion of passengers from auto to high speed rail will result in 1.4 million fewer auto trips in 2010.
- The diversion of passengers from air to high speed rail will result in 60 thousand fewer aircraft flights in 2010.
- Florida high speed rail will serve 1.116 billion passenger miles of travel in 2010.
- An average of 5,380 person-years of employment will be created and supported over the life of the high speed rail franchise.

- During the four peak construction years, the project will increase economic activity by $1.667 billion (1997 dollars) per year in Florida.[1]
- A traveler shifting from auto to high speed rail from Tampa to Miami could be expected to save 2.7 hours of travel time per trip.
- An air or auto traveler shifting to high speed rail between Miami and Orlando in 2010 would be expected to reduce pollutants by 80 pounds and reduce energy consumption by the equivalent of 4.7 gallons of gasoline.
- Based on the expected shifts of demand, the Florida high speed rail project would be expected to prevent 389 auto accidents, 380 auto accident injuries and 5 auto fatalities annually.

This analysis employed the most sophisticated economic modeling tools currently available. Two economic impact assessment models were initially used to determine the economic impacts on the state of Florida of implementing the Florida high speed rail project. The first model was the U.S. Department of Commerce, Regional Input Output Modeling System (RIMS II) which is a static model.[2] The other model is a dynamic, integrated input output econometric model, Regional Economic Model, Inc. (REMI).[3] The REMI model, used in this report, is capable of measuring the socio-economic impacts of variations of economic flows over extended periods of time. This model is widely accepted and has been used extensively over two decades by private, public and academic researchers to simulate the economic impacts of investment and policy options.

The analysis also incorporated financial data and ridership estimates produced and provided to the researchers by the Florida Department of Transportation and FLORIDA OVERLAND EXPRESS (FOX) team members.[4] The projected ridership, costs and revenues for the system are major factors in evaluating the viability of the high speed rail project. Consequently, over the life of the project, changes in fare revenues or costs of operation could affect the resultant financial feasibility and the subsequent economic benefits of the high speed rail project.[5] Similarly, attaining the projected number of riders is a prerequisite to realizing the travel time and congestion reduction benefits estimated in this report.[6] Other benefit calculations, such as safety improvements, energy savings and air quality benefits, are dependent on the ridership forecasts and a variety of estimates of the expected performance of high speed rail and competing modes, over time.

A number of technical, financial, ridership and economic feasibility studies completed over the last decade established the feasibility of high speed rail in the rapidly urbanizing corridors of central and south Florida. These studies also served as a backdrop and foundation for the current analysis. The evaluation in this current study is a melding of the most recent ridership, construction and operation costs, projected revenues, and other financial data into a comprehensive economic impact assessment.

8.2 BACKGROUND

8.2.1 Investments in Transportation

Transportation fulfills many social needs and is considered an essential component of the infrastructure of today's society. Investments in transportation represent significant economic benefits to the community through the movement of people and goods. These benefits accrue directly to those who use the investment as well as indirectly to those who may not use a particular transportation facility. Some benefits, such as the economic stimulus and employment increases from construction, are a direct result of the decision to invest in Florida high speed rail. Other benefits, such as savings in energy, travel time, safety and emissions, are dependent on the ridership and the actual performance of high speed rail in comparison to alternative forms of transport over the life of the project. These benefits are determined based on the best current estimates of ridership and the performance characteristics of air, auto and high speed rail travel.

8.2.2 Growth and Limitations

Florida has experienced population and tourism growth over the past few decades virtually unrivaled elsewhere in the United States. Considerable progress has been made in expanding Florida's highways, ports, airports and public transportation systems; however, growing demand has continued to outpace the supply of new transportation capacity. In addition, it is becoming increasingly clear that the costs and consequences of unlimited expansion of Florida's roadways are more than can be borne by our environment and by the taxpayers.

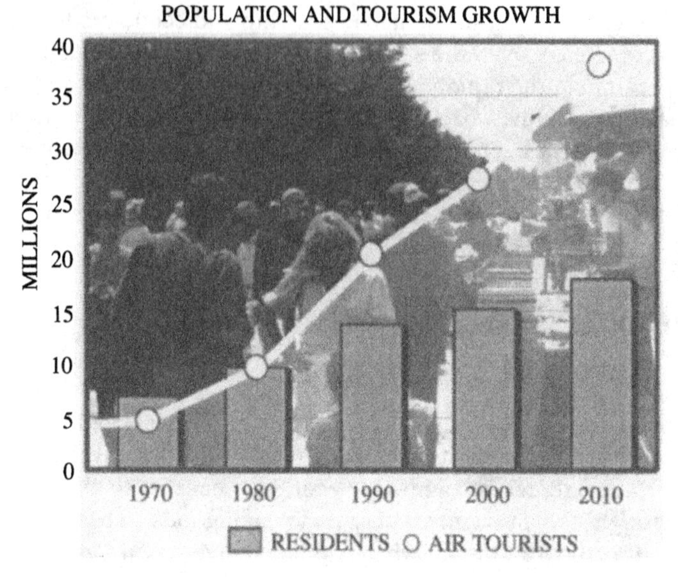

Florida's population grew by 91% between 1970 and 1990 and is projected to increase by 38% between 1990 and 2010. Florida's population growth rate is roughly twice the national population growth rate, air tourist growth is projected to increase by 82% over the same 1990 to 2010 time period.

8.2.3 High Speed Rail Alternative

The Florida Department of Transportation has aggressively sought alternatives to meet the travel needs of Florida residents and tourists while still being responsible stewards of the environment and public resources. In this search, the prospect of implementing a high speed rail system for Florida began in 1982 and is currently mandated by the 1992 Florida High Speed Rail Transportation Act.

Meanwhile, high speed rail has grown more attractive as modern rail technology has proven itself in an increasing number of travel markets across the globe. Florida's rapid population and tourism growth, flat topography, cluster of large urbanized areas, and growing densities have created a travel market that, in part, may best be served by a transportation system that includes high speed rail.

8.2.4 Part of an Integrated Multimodal System

The proposed Florida high speed rail project is not a single cure-all for the pressing travel congestion problems facing the state. High

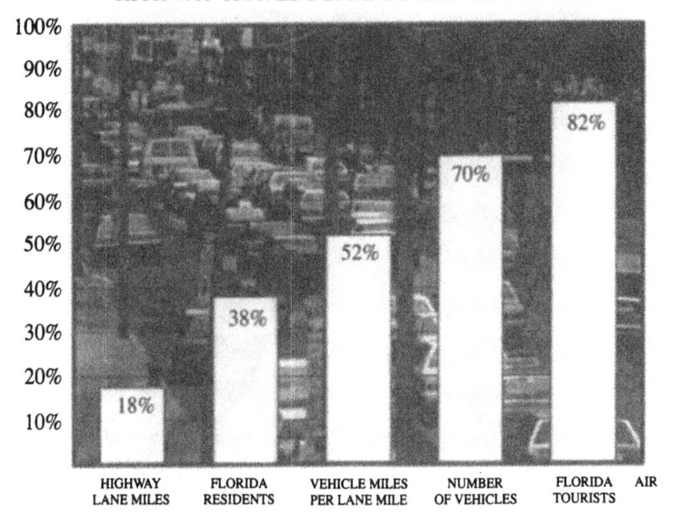

speed rail is, however, one of several pivotal transportation investments needed within the integrated infrastructure of the state to resolve these growing concerns. The Florida high speed rail project is designed to complement other modes of travel. It will serve as an important link in the United States' first multimodal transportation system that will include high speed rail integrated with auto, air travel, buses, vans, park-n-ride facilities, urban rail and other regional transit services.

8.3 FLORIDA HIGH SPEED RAIL PROJECT

In 1996, the Florida Department of Transportation entered into a public/private partnership agreement with FLORIDA OVERLAND EXPRESS (FOX), a consortium of four of the world's largest and most respected international engineering, construction and rail equipment companies, to implement a high speed rail system linking Tampa, Orlando and Miami. The Florida Department of Transportation and FOX are currently in the process of conducting comprehensive studies of ridership, route alignment, construction costs and financing.

The Florida high speed rail system is designed to operate on 320 miles of new electrified track connecting Florida's largest urban areas.

Florida High Speed Rail Route

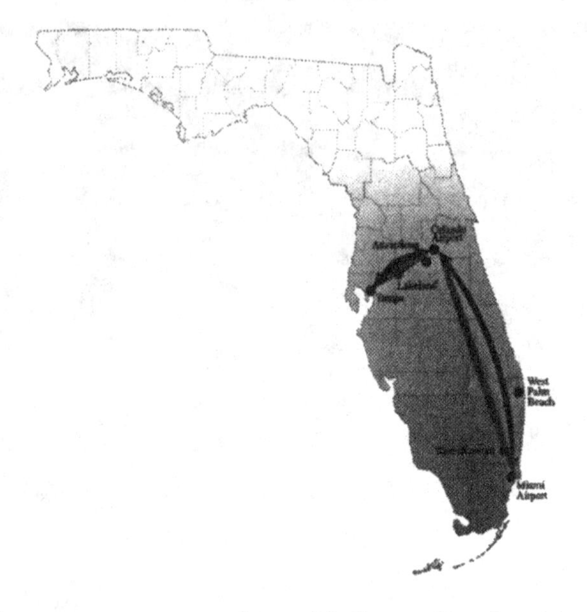

The system proposes connections with five major airports, the highway system, and growing regional rail and bus transit systems across the state's largest metropolitan areas. The counties directly served by this proposed high speed rail system are forecast to contain more than 45% of the state's 15.4 million people by the year 2000 and receive more than 58% of Florida's tourist development tax revenues. Major tourism attractions include recreation areas, beaches, cruise ships, theme parks and urban centers.[9]

Environmental, planning and engineering studies will continue through 1999 and construction is slated to begin in 2000. The first passengers will be able to travel from Miami to Orlando beginning in 2004 and to Tampa in 2006.

8.4 TRANSPORTATION BENEFITS OF HIGH SPEED RAIL

As a precursor to estimating the economic impacts, this study looked at the transportation benefits expected from the project. These benefits are of interest both because they contribute to economic impacts, and because safety, air quality and energy use are among the important considerations in making transportation investment decisions.

Transportation benefits accrue to persons choosing to use high speed rail and to non-users that benefit from the presence of this transportation alternative. These benefits take two forms. The first is the benefit to the traveler, above and beyond the cost of the fare, including consumer surplus, safety, environmental and other savings. The second is the economic and other savings for travelers using existing transportation modes in the form of reductions in congestion as a result of some air and auto travelers switching to high speed rail.

8.4.1 High Speed Rail Ridership

Florida high speed rail is projected to carry 6.13 million one- way riders in the year 2010. This will result in approximately 16,780 daily trips, averaging 182 miles. Forty-six percent of the ridership will be concentrated in the Orlando-Miami segment, with 36 percent and 18 percent in the Tampa-Orlando and Tampa-Miami segments, respectively. Fifty-seven percent of these trips would be made for business purposes, the remainder being personal travel and tourism. Of the total ridership, 31 percent is estimated to shift to high speed rail from air travel, 45 percent from auto, and 24 percent would be new trips induced by the cost and convenience of high speed rail. Approximately 5 per cent of highway traffic between the cities served is expected to shift to high speed rail, while approximately 80 percent of intra-Florida air traffic will be diverted to high speed rail. High speed rail ridership represents about 11 percent of the total travel that starts and ends in the cities served in the Tampa-Orlando-Miami corridor. The average fare is projected to be approximately $64 per trip or $0.35 per passenger mile.[10]

High speed rail would serve approximately 1.1 billion passenger miles of travel in 2010, helping to meet needs in a state that currently has over 127 billion vehicle miles of travel on roadways. These statistics demonstrate how high speed rail would provide significant transportation capacity and carry a large ridership, yet in the context of the total travel demand in Florida, its role, like that of any single project, is to serve as one part of an overall integrated transportation system.

8.4.2 Traveler Benefits

Traveler benefits are estimated by comparing the forecasted performance of the proposed system with the forecasted performance of auto and air modes. Estimates are developed from the ridership forecast (and source of travelers, i.e., shift from air, shift from auto, and induced) and the comparative performance of the modes.[11] Benefits to non-users in the form of time savings from reduced congestion were

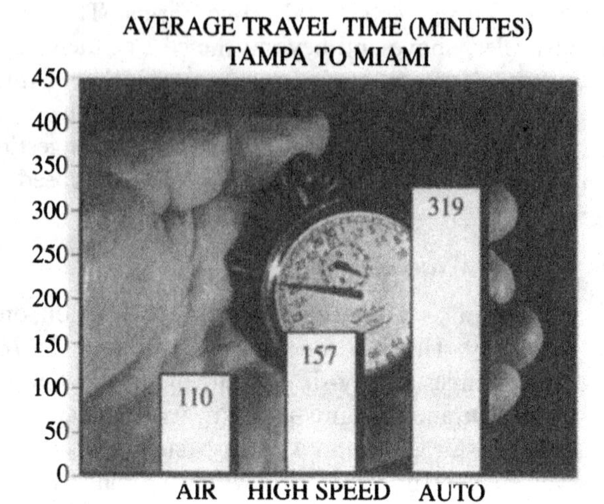

AVOIDED AUTO FATALITIES, INJURIES, AND ACCIDENTS YEARS 2010 AND 2035

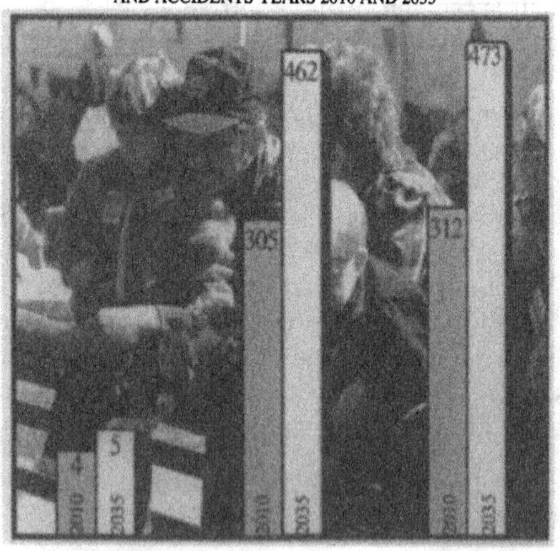

AVOIDED FARTALITIES
AVOIDED INJURIES
AVOIDED ACCIDENTS

ENERGY SAVINGS FROM DIVERTED AUTO AND AIR TRIPS YEARS 2010 AND 2035

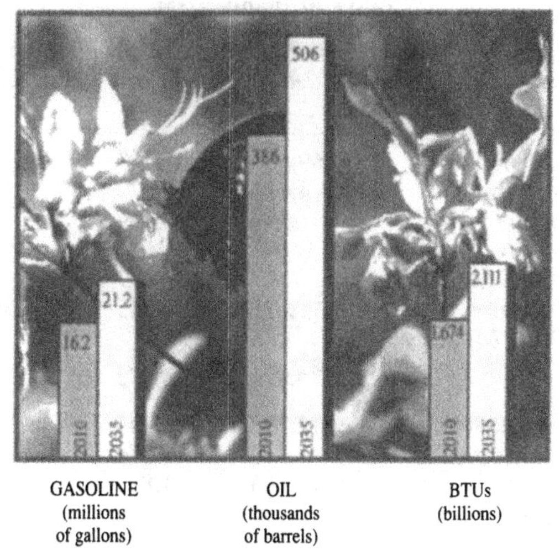

also estimated and included among the inputs to the economic impact assessment.

TIME—High speed rail travelers who shift from auto and air can be expected to save an annual average of 3.8 million hours of time over the 2004 to 2043 time period. The time savings reflect the sum of the estimated door-to-door travel time differences.

SAFETY—Internationally, high speed rail has attained an exceptional safety record which is assumed to continue in Florida operations. Historical trends in air and auto safety are used as a basis for determining the number of accidents, injuries, and fatalities that might be avoided by the shifts of travelers to high speed rail.

ENERGY—Based on forecasted 2010 and 2035 conditions, the presence of the high speed rail service should reduce transportation energy consumption significantly. This savings is the equivalent of 16.2 million gallons of gasoline or 386,000 barrels of oil or 1.674 billion British thermal units in 2010; and 21.2 million gallons of gasoline or 506,000 barrels of oil or 2.111 British thermal units in 2035.

AIR QUALITY—High speed rail will provide reduced pollutants due to travelers shifting from either auto or air travel. The following chart presents changes in tons of pollutants for 2010 and 2035 based on estimated modal characteristics for that time.

TABLE 8.1. Net Reduction in tons of Air Quality Pollutants (Operating Years 2010 and 2035)

Year	*Pollutant*	*Auto*	*Air*	*FHSR**	*Net Reduction*
2010	Carbon Dioxide	69,658	65,260	41,257	93,661
	Carbon Monoxide	4,414	17,220	9	21,625
	Hydrocarbons	595	13,499	1	14,093
	Nitrous Oxides	307	654	191	770
	Particulate Matter	37	97	33	101
	Sulfur Oxides	25	145	287	(117)
	Tire Wear Matter	38			38
Totals		75,074	96,875	41,778	130,171
2035	Carbon Dioxide	112,765	105,645	66,789	151,621
	Carbon Monoxide	7,145	27,876	15	35,006
	Hydrocarbons	963	21,853	2	22,814
	Nitrous Oxides	497	1,058	309	1,246
	Particulate Matter	61	157	54	164
	Sulfur Oxides	40	235	465	(190)
	Tire Wear Matter	62			62
Totals		121,533	156,824	67,634	210,723

*Florida High Speed Rail

8.5 ECONOMIC IMPACTS

8.5.1 Methodology

The basic assumption of this economic impact analysis is that benefits flow from improvements in transportation systems (e.g., reduced travel times and associated costs) and from new dollars attracted to an area. Thus, simply taking taxpayer or private sector dollars and spending them on transportation as opposed to an alternative use will not necessarily create the same positive economic impacts. Realizing positive benefits requires the inflow of new resources (private sector and federal investment in an economic activity) and/or the realization of transportation benefits to travelers that result in savings produced by the new investment.

This analysis evaluates the expected direct and indirect changes in employment, income, and business activity that would be attributable to constructing and operating the high speed rail project in Florida. These impacts result from new money stimulants to the Florida economy that would otherwise not exist in the state's future. The economic impacts measured in this report are only a portion of the total economic benefits that can accrue to Florida from this investment.

8.5.2 Inflow of New Money as Economic Stimulus

The decision to implement a high speed rail project in Florida results in flows of money into the state's economy. The analysis of economic impacts requires an understanding of the flows of money and the consequences that they have on the economy of Florida and the affected regions. A project of this magnitude will have a complex interaction with the state's economy. Investment funds come from several sources: the state, the federal government and the private sector. The principal effects of this investment come from new funds entering the state and from the economic benefits associated with the transportation services that are provided. Thus, the private sector equity investment by the FOX consortium, the contribution of federal funds and the direct benefits associated with improved transportation are the stimulus for additional economic growth. The Florida high speed rail project is forecast to generate a direct return to the state of Florida on its initial and ongoing investment. That investment is secured by state ownership of the high speed rail infrastructure. For the purpose of this analysis, the financial return to the state of Florida is assumed to be reinvested in other high speed rail projects in Florida.

8.5.3 Statewide Employment

Results of the analysis indicate that an estimated 78,102 jobs[12] will be created by the high speed rail project during the planning and construction phase (8 years). An additional 174,786 jobs will be created over the period of operation and reinvestment (39 years). These jobs would be distributed broadly across most sectors of the economy. Out of the average annual 5,380 jobs, the strongest job growth, at approximately 4,000 jobs per year, is in the non-manufacturing sector. This includes growth in services including tourism, transportation, retail trade, finance and government. Additionally, an average annual increase of 200 jobs will be realized in the manufacturing sector during the years of high speed rail operation.

This overall increase in employment is attributable to the construction and operation of the high speed rail system and subsequent economic impacts. The secondary impacts result from the increase in the competitiveness of Florida's businesses and the attractiveness of Florida's economy.

8.5.4 Regional Impacts

Regional impacts are presented for five regions along the route. The study provides estimates of the number of new jobs, annual wages and salaries, and economic activity created by the project. New jobs include those directly involved in the construction and operation of the system, and those that result from stimulated economic activity. Along with the creation of jobs, the project contributes to the economic activity of the state and accordingly increases earnings by the workforce throughout the state.

The projected employment growth is generally distributed in proportion to the spending in each region of the state. This spending will be influenced by the amount of track in the various regions and other infrastructure to be built and operated, such as stations and maintenance facilities. Other employment growth will result from commercial ventures that will support the communities, riders and suppliers affected by the high speed rail system.

In addition, over the long term, funds paid to the state of Florida as return on its investment are available as new revenue for programming and expenditure statewide, as determined by the Florida appropriations process.

Conclusions

The Florida High Speed Rail Project will:
Benefit travelers by:

- Reducing yearly highway travel by over 261 million vechile miles, thus removing 1.4 million auto trips from roads 2010
- Reducing yearly air travel by over 559 million passenger miles, thus enabling a reduction of 60 thousand air craft flights annually in 2010
- Reducing time spent in roadway congestion by over 1.6 million hours per year
- Reducing yearly deaths and injuries by auto travel by 5 and 380, respectively

Benefit the environment by:

- Reducing annual fuel consumption by the equivalent of 16.2 millionk gallons of gas in 2010
- Reducing annual pollutants by 130,171 tons in 2010

Improve florida's economy by:

- Creating 78,102 full time jobs and $2.84 billion in wages and salaries over the first 8 years of the project (planning and construction phase)
- Creating 174,786 full time jobs and $6.04 billion in wages and salaries over the next 39 years. (operations and reinvestment phase)

Provide a stimulus for development of new industries in Florida by:

- Motivating economic development and growth management activities
- Attracting new business and additional tourists to Florida
- Demonstrating a willingness to invest in new ideas, and
- Being a model of public/private partnership.

REGIONAL IMPACTS

These charts illustrate the projected jobs, wages and salaries, and economic activity by region during the construction and operations phases as well as impacts resulting from the reinvestment of surplus revenue.[13]

REGION TOTALS	JOBS	WAGES	ECON. ACT.
		millions of $1	
TAMPA BAY	23,136	$731	$3,166
EAST CENTRAL	55,681	$1,783	$8,261
PALM BEACH CO. & TREASURE COAST	44,253	$1,283	$7,571
BROWARD/DADE COS.	30,777	$1,224	$5,092
OTHER REGIONS	90,043	$3,858	$10,946
TOTAL FLORIDA	252,490	$8,881	$35,037

Impact due to construction and operation of high speed rail

Additional impact due to reinvestment of earnings

JOBS/PERSON YEARS OF EMPLOYMENT

JOBS (ONE PERSON EMPLOYED FULL TIME FOR ONE YEAR)

100,000 90,000 80,000 70,000 60,000 50,000 40,000 30,000 20,000 10,000 0

TAMPA BAY · EAST CENTRAL · PALM BEACH CO. & TREASURE COAST · BROWARD/DADE COS. · OTHER REGIONS

ECONOMIC ACTIVITY

WAGES/SALARIES

THE RESEARCHERS

CEFA Center for Economic Forecasting and Analysis
The Center for Urban Transportation Research (CUTR)

The Florida State University Center for Economic Forecasting and Analysis (CEFA) specializes in applying advanced computer-based economic evaluation and forecasting models that examine and help resolve pressing public policy issues confronting the state of Florida. The Center conducts applied economic research and educational training across a wide range of public policy areas. CEFA staff has decades of applied research leadership in a number of important public and private policy areas. They include expertise in public transportation (with specialized knowledge in the areas of high speed rail and magnetic levitation technology assessment); health care finance and policy; taxation and fiscal issues; health and property insurance; and environmental, land use, energy and planning issues.

TIM LYNCH Dr. Lynch has published internationally on the financing, economic impacts, planning and environmental impacts of high speed rail and magnetic levitation technologies. These works include *The Economics and Financing of High Speed Rail and Maglev Systems in Europe-An Assessment of Financing Methods and Results with the Growing Importance of Public-Private Partnerships and Implications for the U.S.*, National Urban Transit Institute, Florida State University, March 15, 1995, NTIS publication. Dr. Lynch was also the Principal Investigator and Co-authored *An Analysis of the Economic Impacts of Florida High Speed Rail* and *Travel Time, Safety, Energy and Air Quality Impacts of Florida High Speed Rail*, Florida State University and The University of South Florida, July 2, 1997. Dr. Lynch also served on the U.S. Senate Magnetic Levitation Technical Advisory Committee and the U.S. Congressional Technical Clean Air Act Advisory Committee. Dr. Lynch also served as a member of a three National Academy of Science High Speed Ground Transport

Technical Advisory Committees and works and published extensively in the environmental, planning and health care insurance and other economics research areas.

NEIL G. SIPE is a planning consultant who is based in Tallahassee, Florida. Neil specializes in fiscal impact and economic analyses. He has a Ph.D. in urban and regional planning from Florida State University, a M.A. in urban and regional planning from the University of Florida, and a B.A. in environmental studies from New College.

Dr. Sipe worked at the University of Florida's Bureau of Economic & Business Research as a researcher where he specialized in developing microcomputer software for local government economic and financial analysis. He also provided technical assistance to local governments in the areas of planning, economics, and public finance. Additionally, he worked as an economic consultant for an investment banking company based in Central Florida where he conducted feasibility analyses for municipal bond issues and where he established financing districts for residential and mixed-use development projects.

At the University of South Florida conducts a broad range of policy research addressing local, state, and national transportation issues. With a multidisciplinary staff of nearly 40 full-time researchers and 20 graduate assistants, CUTR conducts more than $5 million in sponsored research annually. Since its establishment in 1988 by the Florida legislature, CUTR has completed in excess of 200 projects, valued at nearly $30 million. CUTR houses the National Urban Transit Institute and the editorial office of the Journal of Public Transportation with its distinguished Editorial Board. In the nine years since its inception, CUTR has become one of the top-ranked transportation research centers in the country.

STEVEN E. POLZIN, PE. Deputy Director for Policy Analysis – Center for Urban Transportation Research. Ph.D, Civil Engineering, Transportation, Northwestern University, 1986; MSCE, Urban Systems Engineering, Northwestern, 1976; BSCE, Civil and Environmental Engineering, University of Wisconsin, Madison, 1974. Specialities: public transportation, public policy analysis, transportation planning, systems evaluation, planning process design, mobility analysis. Professional Activities: TRB A1E12, Light Rail Transit, member.

XUEHAO CHU. Research Associate – Center for Urban Transportation Research. Ph.D, Economics, University of California at Irvine, 1993; MA, Economics, University of California at Irvine, 1991; BS, Mathematics, Hangzhou University, China, 1982. Specialities: transportation economics, urban and regional economics, discrete choice analysis, quantitative methods.

ENDNOTES

1 All references to dollars in this report are in 1997 dollars.
2 *Regional Multipliers: A Users Handbook for the Regional Input-Output Modeling System (RIMSII)*, U.S. Department of Commerce, Economics and Statistics, 1992.
3 Treyz, George I., *Regional Economic Modeling: A Systematic Approach to Economic Forecasting and Policy Analysis*, Norwell, MA: Kluwer Academic Publishers, 1993.
4 The information provided to the researchers by Florida Department of Transportation and FOX was based on actual quantity of material estimates, locally-based unit prices of material, labor to install, staffing of construction and operational phases, and maintenance of the system during the franchise per Florida Department of Transportation franchise agreements: FOX Proposal dated October 31, 1995; Post-Franchise Agreement dated August 2, 1996; Pre-Certification Post-Franchise Agreement dated November 12, 1996.
5 FOX will be implementing maximum yield pricing as currently employed by the airline industry and other transportation providers.
6 In addition to the 1993 Florida Department of Transportation ridership study forecast incorporated in the FOX proposal, two independent ridership studies are currently being conducted.
7 *Consensus Forecast*, Florida Demographic and Economic Research, Florida Legislature, 1997.
8 Extrapolation of Florida Department of Transportation, Bureau of Transportation Statistics data.
9 Floyd, Susan S., *1996 Florida Statistical Abstract, 13th Edition*, Gainesville, FL: University of Florida, 1996
10 Florida Department of Transportation and FOX Pre- Certification Post-Franchise Agreement dated November 12, 1996, and supporting documents.
11 Lynch, T. and Polzin, S., *Travel Time, Safety, Energy and Air Quality Impacts of Florida High Speed Rail*, 1997. Numbers have been rounded.
12 According to accepted industry standards, a job is defined as one person employed full time for one year.
13 Lynch, T. and Polzin, S., *An Analysis of the Economic Impacts of Florida High Speed Rail*, 1997. Numbers have been rounded.

Printed in the United States
by Baker & Taylor Publisher Services